Are the Planets Inhabited?

By

E. Walter Maunder

tion et mise en page : Culturea (Hérault, 34)
trouvez notre catalogue sur http://culturea.fr/
ntact : infos@culturea.fr
primé en Allemagne par Books on Demand Gmbh
sign typographique et layout : Derek Murphy et Reedsy
BN :9791041985838
te de parution : février 2024

CHAPTER I

THE QUESTION STATED

The first thought that men had concerning the heavenly bodies was an obvious one: they were lights. There was a greater light to rule the day; a lesser light to rule the night; and there were the stars also.

In those days there seemed an immense difference between the earth upon which men stood, and the bright objects that shone down upon it from the heavens above. The earth seemed to be vast, dark, and motionless; the celestial lights seemed to be small, and moved, and shone. The earth was then regarded as the fixed centre of the universe, but the Copernican theory has since deprived it of this pride of place. Yet from another point of view the new conception of its position involves a promotion, since the earth itself is now regarded as a heavenly body of the same order as some of those which shine down upon us. It is amongst them, and it too moves and shines—shines, as some of them do, by reflecting the light of the sun. Could we transport ourselves to a neighbouring world, the earth would seem a star, not distinguishable in kind from the rest.

But as men realized this, they began to ask: "Since this world from a distant standpoint must appear as a star, would not a star, if we could get near enough to it, show itself also as a world? This world teems with life; above all, it is the home of human life. Men and women, gifted with feeling, intelligence, and character, look upward from its surface and watch the shining members of the heavenly host. Are none of these the home of beings gifted with like powers, who watch in their turn the movements of that shining point which is our world?"

This is the meaning of the controversy on the Plurality of Worlds which excited so much interest some sixty years ago, and has been with us more or less ever since. It is the desire to recognize the presence in the orbs around us of beings like ourselves, possessed of personality and intelligence, lodged in an organic body.

This is what is meant when we speak of a world being "inhabited." It would not, for example, at all content us if we could ascertain that

Jupiter was covered by a shoreless ocean, rich in every variety of fish; or that the hard rocks of the Moon were delicately veiled by lichens. Just as no richness of vegetation and no fulness and complexity of animal life would justify an explorer in describing some land that he had discovered as being "inhabited" if no men were there, so we cannot rightly speak of any other world as being "inhabited" if it is not the home of intelligent life. If the life did not rise above the level of algæ or oysters, the globe on which they flourish would be uninhabited in our estimation, and its chief interest would lie in the possibility that in the course of ages life might change its forms and develop hereafter into manifestations with which we could claim a nearer kinship.

On the other hand, of necessity we are precluded from extending our enquiry to the case of disembodied intelligences, if such be conceived possible. All created existences must be conditioned, but if we have no knowledge of what those conditions may be, or means for attaining such knowledge, we cannot discuss them. Nothing can be affirmed, nothing denied, concerning the possibility of intelligences existing on the Moon or even in the Sun if we are unable to ascertain under what limitations those particular intelligences subsist. Gnomes, sylphs, elves, and fairies, and all similar conceptions, escape the possibility of discussion by our ignorance of their properties. As nothing can be asserted of them they remain beyond investigation, as they are beyond sight and touch.

The only beings, then, the presence of which would justify us in regarding another world as "inhabited" are such as would justify us in applying that term to a part of our own world. They must possess intelligence and consciousness on the one hand; on the other, they must likewise have corporeal form. True, the form might be imagined as different from that we possess; but, as with ourselves, the intelligent spirit must be lodged in and expressed by a living material body. Our enquiry is thus rendered a physical one; it is the necessities of the living body that must guide us in it; a world unsuited for living organisms is not, in the sense of this enquiry, a "habitable" world.

The discussion, as it was carried on sixty years ago by Dr. Whewell and Sir David Brewster, was essentially a metaphysical, almost a theological one, and it was chiefly considered in its supposed relationship to certain religious conceptions. It was urged that it was derogatory to the wisdom and goodness of the Creator to suppose that He would have created so many great and glorious orbs without having a definite purpose in so doing, and that the only purpose for which a world could be made was that it might be inhabited. So, again, when Dr. A. R. Wallace revived the discussion in 1903, he clearly had a theological purpose in his opening paper, though he was taking the opposite view from that held by Brewster half a century earlier.

[Pg 5]For myself, if there be any theological significance attaching to the solving of this problem, I do not know what it is. If we decide that there are very many inhabited worlds, or that there are few, or that there is but one—our own—I fail to see how it should modify our religious beliefs. For example: explorers have made their way across the Antarctic continent to the South Pole but have found no "inhabitant" there. Has this fact any theological bearing? or if, on the contrary, a race of men had been discovered there, what change would it have made in the theological position of anyone? And if this be so with regard to a new continent on this earth, why should it be different with regard to the continents of another planet?

The problem therefore seems not to be theological or metaphysical, but purely physical. We have simply to ask with regard to each heavenly body which we pass in review: "Are its physical conditions, so far as we can ascertain them, such as would render the maintenance of life possible upon it?" The question is not at all as to how life is generated on a world, but as to whether, if once in action on a particular world, its activities could be carried on.

CHAPTER II

THE LIVING ORGANISM

A world for habitation, then, is a world whereon living organisms can exist that are comparable in intelligence with men. But "men" presuppose the existence of living organisms of inferior grades. Therefore a world for habitation must first of all be one upon which it is possible for living organisms, as such, to exist.

It does not concern us in the present connection how life first came into existence on this planet. It is sufficient that we know from experience that life does exist here; and in whatsoever way it was first generated here, in that same way we may consider that it could have been generated on another planet.

Nor need any question trouble us as to the precise line of demarkation to be drawn between inorganic and organic substances, or amongst the latter, between plants and animals. These are important subjects for discussion, but they do not affect us here, for we are essentially concerned with the highest form of organism, the one furthest from these two dividing lines.

[Pg 7]It suffices that living organisms do exist here, and exist under well-defined conditions. Wanting these conditions, they perish. We can, to a varying degree, determine the physical conditions prevailing upon the heavenly bodies, and we can ascertain whether these physical conditions would be favourable, unfavourable, or fatal to the living organism.

What is a living organism? A living organism is such that, though it is continually changing its substance, its identity, as a whole, remains essentially the same. This definition is incomplete, but it gives us a first essential approximation, it indicates the continuance of the whole, with the unceasing change of the details. Were this definition complete, a river would furnish us with a perfect example of a living organism, because, while the river remains, the individual drops of water are continually changing. There is then something more in the living organism than the continuity of the whole, with the change of the details.

An analogy, given by Max Verworn, carries us a step further. He likens life to a flame, and takes a gas flame with its butterfly shape as a particularly appropriate illustration. Here the shape of the flame remains constant, even in its details. Immediately above the burner, at the base of the flame, there is a completely dark space; surrounding this, a bluish zone that is faintly luminous;[Pg 8] and beyond this again, the broad spread of the two wings that are brightly luminous. The flame, like the river, preserves its identity of form, while its constituent details—the gases that feed it—are in continual change. But there is not only a change of material in the flame; there is a change of condition. Everywhere the gas from the burner is entering into energetic combination with the oxygen of the air, with evolution of light and heat. There is change in the constituent particles as well as change of the constituent particles; there is more than the mere flux of material through the form; there is change of the material, and in the process of that change energy is developed.

A steam-engine may afford us a third illustration. Here fresh material is continually being introduced into the engine there to suffer change. Part is supplied as fuel to the fire there to maintain the temperature of the engine; so far the illustration is analogous to that of the gas flame. But the engine carries us a step further, for part of the material supplied to it is water, which is converted into steam by the heat of the fire, and from the expansion of the steam the energy sought from the machine is derived. Here again we have change in the material with development of energy; but there is not only work done in the subject, there is work done by it.

But the living organism differs from artificial machines in that, of itself and by itself, it is [Pg 9]continuously drawing into itself non-living matter, converting it into an integral part of the organism, and so endowing it with the qualities of life. And from this non-living matter it derives fresh energy for the carrying on of the life of the organism.

The engine and the butterfly gas flame do not give us, any more than the river, a complete picture of the living organism. The form of the river is imposed upon it from without; the river is defined by its bed, by the contour of the country through which it flows. The form and size of the flame are equally defined by exterior conditions; they are imposed upon it by the shape of the burner and the pressure of the gas passing through it. The form of the engine is as its designer has made it. But the form of the living organism is imposed upon it from within; and, as far as we can tell, is inherent in it. Here is the wonder and mystery of life: the power of the living organism to assimilate dead matter, to give it life and bring it into the law and unity of the organism itself. But it cannot do this indiscriminately; it is not able thus to convert every dead material; it is restricted, narrowly restricted, in its action. "One of the chief characteristics of living matter is found in the continuous range of chemical reactions which take place between living cells and their inorganic surroundings. Without cease certain substances are taken up and disappear in the endless round of chemical reactions in the cell. Other substances which have been produced by the chemical reactions in living matter pass out of the cell and reappear in inorganic nature as waste products of the life process. The whole complex of these chemical transformations is generally called *Metabolism*. Inorganic matter contrasts strikingly with living substance. However long a crystal or a piece of metal is kept in observation, there is no change of the substance, and the molecules remain the same and in the same number. For living matter the continuous change of substances is an indispensable condition of existence. To stop the supply of food material for a certain time is sufficient to cause a serious lesion of the life process or even the death of the cell. But the same happens when we hinder the passing out of the products of chemical transformation from the cell. On the other hand, we may keep a crystal of lifeless matter in a glass tube carefully shut up from all exchange of substance with the external world for as many years as we like. The existence of this crystal will continue without end and without change of any of its properties. There is no known living organism which could remain in a dry resting state for an infinitely long period of time. The longest lived are perhaps the spores of mosses which can exist in a dry state more than a hundred years. As a rule the seeds of higher plants show their vital power already weakened after ten years; most of them do not germinate if kept more than twenty to thirty years. These experiences lead to the opinion that even dry seeds and spores of lower plants in their period of rest of vegetation continue the processes of metabolism to a certain degree. This supposition is confirmed by the fact that a very slight respiration and production of carbonic acid can be proved when the seeds contain a small percentage of water. It seems as if life were weakened in these plant organs to a quite imperceptible degree, but never, not even temporarily, really suspended.

"Life is, therefore, quite inseparable from chemical reactions, and on the whole what we call life is nothing else but a complex of innumerable chemical reactions in the living substance which we call protoplasm."

The essential quality, therefore, of life is continual change, but not mere change in general. It is that special process of the circulation of matter which we call metabolism, and this circulation is always connected with a particular chemical substance—protoplasm.

In this substance five elements are always present and predominant—carbon, oxygen, nitrogen, hydrogen, and sulphur. The compounds which these five elements form with each other are most complex and varied, and they also admit to combination—but in smaller proportions—some of the other elements, of which phosphorus, potassium, calcium, magnesium, and iron are the most important.

For protoplasm—using the term in the most general sense—is a chemical substance, not a mere mixture of a number of chemical elements, nor a mere mechanical structure. "However differently the various plasma substances behave in detail, they always exhibit the same general composition as the other albuminoids out of the five 'organo-genetic elements'—namely in point of weight, 51-54% carbon, 21-23% oxygen, 15-17% nitrogen, 6-7% hydrogen, and 1-2% sulphur."

Haeckel, the writer just quoted, describes the plasm, the universal basis of all the vital phenomena, in the following terms: "In every case where we have with great difficulty succeeded in examining the plasm as far as possible and separating it from the plasma-products, it has the appearance of a colourless, viscous substance, the chief physical property of which is its peculiar thickness and consistency. The physicist distinguishes three conditions of inorganic matter—solid, fluid, and gaseous. Active living protoplasm cannot be strictly described as either fluid or solid in the physical sense. It presents an intermediate stage between the two which is best described as viscous; it is best compared to a cold jelly, or solution of glue. Just as we find the latter substance in all stages between the solid and the fluid, so we find in the case of protoplasm. The cause of this softness is the quantity of water contained in the living matter, which generally amounts to a half of its volume and weight. The water is distributed between the plasma molecules or the ultimate particles of living matter in much the same way as it is in the crystals of salts, but with the important difference that it is very variable in quantity in the plasm. On this depends the capacity for the absorption or imbibition in the plasm, and the mobility of its molecules, which is very important for the performance of the vital actions. However, this capacity of absorption has definite limits in each variety of plasm; living plasm is not soluble in water, but absolutely resists the penetration of any water beyond this limit." And Czapek further tells us that "the most striking feature of cell life is the fact that an enormous number of chemical reactions take place within the narrowest space. Most plant cells do not exceed 0·1 to 0·5 millimetres in diameter. Their greatest volume therefore can only be an eighth of a cubic millimetre. Nevertheless, in this minute space we notice in every stage of cell life a considerable number of chemical reactions which[Pg 14] are carried on contemporaneously, without one disturbing the other in the slightest degree."

It is clear if organic bodies were built up of chemical compounds of small complexity and great stability that this continuous range of chemical reactions, this unceasing metabolism, could not take place. It is therefore a necessary condition for organic substances that they should be built up of chemical compounds that are most complex and unstable. "Exactly those substances which are most important for life possess a very high molecular weight, and consequently very large molecules, in comparison with inorganic matter. For example: egg-albumin is said to have the molecular weight of at least 15,000, starch more than 30,000, whilst the molecular weight of hydrogen is 2, of sulphuric acid and of potassium nitrate about 100, and the molecular weight of the heaviest metal salts does not exceed about 300."

To sum up: the living organism, whether it be a simple cell, or the ordered community of cells making up the perfect plant or animal, is an entity, a living individual, wherein highly complex and unstable compounds are unceasingly undergoing chemical reactions, a metabolism essentially associated with protoplasm. But these complex compounds are, nevertheless, formed by the combinations of but a few of the elements now known to us.

Many writers on the subject of the habitability of other worlds, from contemplating the rich and apparently limitless variety of the forms of life, and the diversity of the conditions under which they exist, have been led to assume that the basis of life must itself also in like manner be infinitely broad and infinitely varied. In this they are mistaken. As we have seen, the elements entering into the composition of organic bodies are, in the main, few in number. The temperatures at which they can exist are likewise strictly limited. But, above all, that circulation of matter which we call Life—the metabolism of vital processes—requires for its continuance the presence of one indispensable factor—WATER.

Protoplasm itself, as Czapek puts it, is practically an *albumin sol*; that is to say, it is a chemical substance of which the chief constituents are albuminous matter and water, and the protoplasm can only take from without material dissolved in water; it can only eject matter in the same way. This *osmosis* is an indispensable condition in the vital process. And the "streaming" of protoplasm, its continual movement in the cell, can only be carried on in water.

WATER is the compound of oxygen and hydrogen in the proportion of two atoms of hydrogen to one of oxygen. It is familiar to us in three states: solid, liquid, and gaseous, or ice, water, and steam. But it is only in the liquid state that water is available for carrying on the processes of life. This fact limits the temperatures at which the organic functions can be carried on, for water under terrestrial conditions is only liquid for a hundred degrees; it freezes at 0° Centigrade, it boils at 100° Centigrade. Necessarily, our experiences are mostly confined within this range, and therefore we are apt unconsciously to assume that this range is all the range that is possible, whereas it is but a very small fraction of the range conceivable, and indeed existing, in cosmical space. In its liquid state water is a general solvent, and yet pure water is neutral in its qualities, both characteristics being essential to its usefulness as a vehicle for the protoplasmic actions. Naturally, this function of water as a solvent can only exist when water is in the liquid state; solid water, that is ice, neither dissolves nor flows; and water, when heated to boiling point, passes into vapour, and so leaves the organism moistureless, and therefore dead. It is possible to grind a living organism to a pulp so that the structure of the cells is practically destroyed, and yet for some reactions which are quite peculiar to life still to show themselves for some appreciable time. But when the cell-pulp is heated to the temperature of boiling water, these chemical processes cannot be longer observed. What is left may then be considered as definitely dead.

Water is, then, indispensable for the living organism; but there are two great divisions of such organisms—plants and animals. Animals are generally, but not universally, free to move, and therefore to travel to seek their food. But their food is restricted; they cannot directly convert inorganic matter to their own use; they can only assimilate organic material. The plant, on the other hand, unlike the animal, can make use of inorganic material. Plant life, therefore, requires an abundant supply of water in which the various substances necessary for its support can be dissolved; it must either be in water, or, if on land, there must be an active circulation of water both through the atmosphere and through the soil, so as to bring to it the food that it requires. Animal life presupposes plant life, for it is always dependent upon it.

Many writers have assumed that life is very widely distributed in connection with this planet. The assumption is a mistaken one, as has been well pointed out by Garrett P. Serviss, a charming writer on astronomical subjects: "On the Earth we find animated existence confined to the surface of the crust of the globe, to the lower and denser strata of the atmosphere, and to the film of water that constitutes the oceans. It does not exist in the heart of the rocks forming the body of the planet nor in the void of space surrounding it outside the atmosphere. As the Earth condensed from the original nebula, and cooled and solidified, a certain quantity of matter remained at its surface in the form of free gases and unstable compounds, and, within the narrow precincts where these things were, lying like a thin shell between the huge inert globe of permanently combined elements below, and the equally unchanging realm of the ether above, life, a phenomenon depending upon ceaseless changes, combinations and re-combinations of chemical elements in unstable and temporary union, made its appearance, and there only we find it at the present time."

"The huge inert globe of permanently combined elements below, and the equally unchanging realm of the ether above," offer no home for the living organism; least of all for the highest of such organisms—Man. Both must be tempered to a condition which will permit and favour continual change, the metabolism which is the essential feature of life.

"When the earth had to be prepared for the habitation of man, a veil, as it were, of intermediate being was spread between him and its darkness, in which were joined, in a subdued measure, the stability and the insensibility of the earth, and the passion and perishing of mankind.

"But the heavens, also, had to be prepared for his habitation. Between their burning light,—their deep vacuity, and man, as between the earth's gloom of iron substance, and man, a veil had to be[Pg 19] spread of intermediate being;—which should appease the unendurable glory to the level of human feebleness, and sign the changeless motion of the heavens with the semblance of human vicissitude. Between the earth and man arose the leaf. Between the heaven and man came the cloud. His life being partly as the falling leaf and partly as the flying vapour."

The leaf and the cloud are the signs of a habitable world. The leaf—that is to say, plant life, vegetation—is necessary because animal life is not capable of building itself up from inorganic material. This step must have been previously taken by the plant. The cloud, that is to say water-vapour, is necessary because the plant in its turn cannot directly assimilate to itself the nitrogen from the atmosphere. The food for the plant is brought to it by water, and it assimilates it by the help of water. It is, therefore, upon the question of the presence of water that the question of the habitability of a given world chiefly turns. In the physical sense, man is "born of water," and any world fitted for his habitation must "stand out of the water and in the water."

CHAPTER III

THE SUN

The Sun is, of all the heavenly bodies, the most impressive, and has necessarily, at all times, attracted the chief attention of men. There are only two of the heavenly bodies that appear to be more than points of light, only two that show a surface to the naked eye, and the Sun, being so much the brighter of the two, and the obvious source of all our light and heat, and the fosterer of vegetation, readily takes the premier place in interest. In the present day we know too much about the Sun for anyone to suppose that it can be the home of organic life; but it is not many years since its habitability was seriously suggested even by so high an authority as Sir William Herschel. He conceived that it was possible that its stores of light and heat might be confined to a relatively thin shell in its upper atmosphere, and that below this shell a screen of clouds might so check radiation downward that it would be possible for an inner nucleus to exist which should be cool and solid. This fancied inner globe would then necessarily enjoy perpetual daylight, and a climate which knew no variation from pole to pole. To its inhabitants the entire heavens would be generally luminous, the light not being concentrated into any one part of the vault; and it was supposed that, ignorant of time, a happy race might flourish, cultivating the far-spread solar fields, in perpetual daylight, and in the serenity of a perpetual spring that was distracted by no storm.

The picture thus conjured up is a pleasing one, though probably, to the restless sons of Earth, it would seem to suffer somewhat from monotony. But we now know that it corresponds in not a single detail to the actual facts. The study of solar conditions carried on through the last hundred years has revealed to us, not serenity and peace, but storm, stress, and commotion on the most gigantic scale. But though we now can dismiss from our minds the possibility that the Sun can be inhabited, yet it is of such importance to the maintenance of life on this planet, and by parity of reasoning to life on any other planet, that a review of its conditions forms a necessary introduction to our subject. Further, those conditions themselves will bring out certain principles that are of necessary application when we come to consider the case of particular planets.

The distance of the Sun from the Earth is often spoken of as the "astronomical unit"; it is the fundamental measure of astronomy, and all our information as to the sizes and distances of the various planets rests upon it. And, as we shall shortly see, the particular problem with which we are engaged—the habitability of worlds—is directly connected with these two factors: the size of the world in question, and its distance from the Sun.

The distance of the Sun has been determined by several different methods the principles of which do not concern us here, but they agree in giving the mean distance of the Sun as a little less than 93,000,000 miles; that is to say, it would require 11,720 worlds as large as our own to be put side by side in order to bridge the chasm between the two. Or a traveller going round the Earth at its equator would have to repeat the journey 3730 times before he had traversed a space equal to the Sun's distance.

But knowing the Sun's distance, we are able to deduce its actual diameter, its superficial extent, and its volume, for its apparent diameter can readily be measured. Its actual diameter then comes out as 866,400 miles, or 109·4 times that of the Earth. Its surface exceeds that of the Earth 11,970 times; its volume, 1,310,000 times.

But the weight of the Sun is known as well as its size; this follows as a consequence of gravitation. For the planets move in orbits under the influence of the Sun's attraction; the dimensions of their orbits are known, and the times taken in describing them; the amount of the attractive force therefore is also known, that is to say, the mass of the Sun. This is 332,000 times the mass of the[Pg 23] Earth; and as the latter has been determined as equal to about

6,000,000,000,000,000,000,000 tons

that of the Sun would be equal to

2,000,000,000,000,000,000,000,000,000 tons.

It will be seen that the proportion of the volume of the Sun to that of the Earth is greater than the proportion of its mass to the Earth's mass—almost exactly four times greater; so that the mean density of the Sun can be only one-fourth that of the Earth. Yet, if we calculate the force of gravity at the surfaces of both Sun and Earth, we find that the Sun has a great preponderance. Its mass is 332,000 times that of the Earth, but to compare it with the attraction of the Earth's surface we must divide by $(109{\cdot}4)^2$, since the distance of the Sun's centre from its surface is 109·4 times as great as the corresponding distance in the case of the Earth, and the force of gravity diminishes as the square of the increased distance. This gives the force of gravity at the solar surface as 27·65 times its power at the surface of the Earth, so that a body weighing one ton here would weigh 27 tons 13 cwt. if it were taken to the Sun.

This relation is one of great importance when we realize that the pressure of the Earth's atmosphere[Pg 24] is 14·7 lb. on the square inch at the sea level; that is to say, if we could take a column of air one square inch in section, extending from the surface of the Earth upwards to the very limit of the atmosphere, we should find that it would have this weight. If we construct a water barometer, the column of water required to balance the atmosphere must be 34 feet high, while the height of the column of mercury in a mercurial barometer is 30 inches high, for the weight of 30 cubic inches of mercury or of 408 cubic inches of water (34 × 12 = 408) is 14·7 lb.

If, now, we ascend a mountain, carrying a mercurial barometer with us we should find that it would fall about one inch for the first 900 feet of our ascent; that is to say, we should have left one-thirtieth of the atmosphere below us by ascending 900 feet. As we went up higher we should find that we should have to climb more than 900 feet further in order that the barometer might fall another inch; and each successive inch, as we went upward, would mean a longer climb. At the height of 2760 feet the barometer would have fallen three inches; we should have passed through one-tenth of the atmosphere. At the height of 5800 feet, we should have passed through one-fifth of the atmosphere, the barometer would have dropped six inches; and so on, until at about three and a third miles above sea level the barometer would read fifteen inches, showing that we had passed through half the atmosphere. Mont Blanc is not quite three miles high, so that in Europe we cannot climb to the height where half the atmosphere is left below us, and there is no terrestrial mountain anywhere which would enable us to double the climb; that is to say, to ascend six and two-third miles. Could we do so, however, we should find that the barometer had fallen to seven and a half inches; that the second ascent of three and a third miles had brought us through half the remaining atmosphere, so that only one-fourth still remained above us. In the celebrated balloon ascent made by Mr. Coxwell and Mr. Glaisher on September 5, 1861, an even greater height was attained, and it was estimated that the barometer fell at its lowest reading to seven inches, which would correspond to a height of 39,000 feet.

But on the Sun, where the force of gravity is 27·65 times as great as at the surface of the Earth, it would, if all the other conditions were similar, only be necessary to ascend one furlong, instead of three and a third miles, in order to reach the level of half the surface pressure, and an ascent of two furlongs would bring us to the level of quarter pressure, and so on. If then the solar atmosphere extends inwards, below the apparent surface, it should approximately double in density with each furlong of descent. These considerations, if taken alone, would point to a mean density of the Sun not as we know it to be, less than that of the Earth, but immeasurably greater; but the discordance is sufficiently explained when we come to another class of facts.

These relate to the temperature of the Sun, and to the enormous amount of light and heat which it radiates forth continually. This entirely transcends our power to understand or appreciate. Nevertheless, the astonishing figures which the best authorities give us may, by their vastness, convey some rough general impression that may be of service. Thus Prof. C. A. Young puts the total quantity of sunlight as equivalent to

1,575,000,000,000,000,000,000,000,000 standard candles.

The intensity of sunlight at each point of the Sun's surface is variously expressed as

190,000 times that of a standard candle,
5300 times that of the metal in a Bessemer converter,
146 times that of a calcium light,
or, 3·4 times that of an electric arc.

The same authority estimates at 30 *calories* the value of the *Solar Constant*; that is to say, the heat which, if our atmosphere were removed, would be received from the Sun in a minute of time upon a square metre of the Earth's surface that had the Sun in its zenith, would be sufficient to raise the temperature of a kilogram of water 30 degrees Centigrade. This would involve that the heat radiation from each square metre of the Sun's surface would equal 1,340,000 calories; or sufficient to melt through in each minute of time a shell of ice surrounding the Sun to the thickness of 58·2 feet. Prof. Abbot's most recent determination of the solar constant diminishes these estimates by one third; but he still gives the probable temperature of the solar surface as not far short of 7000 degrees Centigrade, or about 12,000 degrees Fahrenheit.

The Sun, then, presents us with temperatures and pressures which entirely surpass our experience on the Earth. The temperatures, on the one hand, are sufficient to convert into a permanent gas every substance with which we are acquainted; the pressures, on the other hand, apart from the high temperatures, would probably solidify every element, and the Sun, as a whole, would present itself to us as a comparatively small solid globe, with a density like that of platinum. With both factors in operation, we have the result already given: a huge globe, more than one hundred times the diameter of the Earth, yet only one-fourth its density, and gaseous probably throughout the whole of its enormous bulk.

What effect have these two factors, so stupendous in scale, upon its visible surface? What is the appearance of the Sun?

It appears to be a large glowing disc, sensibly circular in outline, with its edge fairly well-defined both as seen in the telescope and as registered on photographs. In the spectroscope, or when in an eclipse of the Sun the Moon covers the whole disc, a narrow serrated ring is seen surrounding the rim, like a velvet pile of a bright rose colour. This crimson rim, the sierra or *chromosphere* as it is usually called, is always to be found edging the entire Sun, and therefore must carpet the surface everywhere. But under ordinary conditions, we do not see the chromosphere itself, but look down through it on the *photosphere*, or general radiating surface. This, to the eye, certainly looks like a definite shell, but some theorists have been so impressed with the difficulty of conceiving that a gaseous body like the Sun could, under the conditions of such stupendous temperatures as there exist, have any defined limit at all, that they deny that what we see on the Sun is a real boundary, and argue that it only appears so to us through the effects of the anomalous refraction or dispersion of light. Such theories introduce difficulties greater and more numerous than those that they clear away, and they are not generally accepted by practical observers of the Sun. They seem incompatible with the apparent structure of the photosphere, which is everywhere made up of a complicated mottling: minute grains somewhat resembling those of rice in shape, of intense brightness, and irregularly scattered. This mottling is sometimes coarsely, sometimes finely textured; in some regions it is sharp and well defined, in others misty or blurred, and in both cases they are often arranged in large elaborate patterns, the figures of the pattern sometimes extending for a hundred thousand miles or more in any direction. The rice-like grains or granules of which these figures are built up, and the darker pores between them, are, on the other hand, comparatively small, and do not, on the average, exceed two to four hundred miles in diameter.

But the Sun shows us other objects of quite a different order in their dimensions. Here and there the bright granules of the photosphere become disturbed and torn apart, and broad areas are exposed which are relatively dark. These are *sunspots*, and in the early stages of their development they are usually arranged in groups which tend to be stretched out parallel to the Sun's equator. A group of spots in its later stages of development is more commonly reduced to a single round, well-defined, dark spot. These groups, when near the edge of the Sun, are usually seen to be accompanied by very bright markings, arranged in long irregular lines, like the foam on an incoming tide. These markings are known as the *faculae*, from their brightness. In the spectroscope, when the serrated edges of the chromosphere are under observation, every now and then great *prominences*, or tongues and clouds of flame, are seen to rise up from them, sometimes changing their form and appearance so rapidly that the motion can almost be followed by the eye. An interval of fifteen or twenty minutes has frequently been sufficient to transform, quite beyond recognition, a mass of flame fifty thousand miles in height. Sometimes a prominence of these, or even greater, dimensions has formed, developed, risen to a great distance from the Sun, and completely disappeared within less than half an hour. The velocity of the gas streams in such eruptions often exceeds one hundred miles a second; sometimes, though only rarely, it reaches a speed twice as great.

Sunspots do not offer us examples of motions of this order of rapidity, but the areas which they affect are not less astonishing. Many spot groups have been seen to extend over a length of one hundred thousand, or one hundred and fifty thousand miles, and to cover a total area of a thousand million square miles. Indeed, the great group of February, 1905, at its greatest extent, covered an area four times as great as this. Again, in the normal course of the development of a spot group, the different members of the group frequently show a kind of repulsion for each other in the early stages of the group's history, and the usual speed with which they move away from each other is three hundred miles an hour.

The spots, the faculae, the prominences, are all, in different ways, of the nature of storms in an atmosphere; that is to say, that, in the great gaseous bulk of the Sun, certain local differences of constitution, temperature, and pressure are marked by these different phenomena. From this point of view it is most significant that many spots are known to last for more than a month; some have been known to endure for even half a year. The nearest analogy which the Earth supplies to these disturbances may be found in tropical cyclones, but these are relatively of far smaller area, and only last a few days at the utmost, while a hundred miles an hour is the greatest velocity they ever exhibit, and this, fortunately, only under exceptional circumstances. For a wind of such violence mows down buildings and trees as a scythe the blades of grass; and were tornadoes moving at a rate of 300 miles an hour as common upon the Earth as spots are upon the Sun, it would be stripped bare of plants and animals, as well as of men and of all their works.

It is not an accident that the Sun, when storm-swept, shows this violence of commotion, but a necessary consequence of its enormous temperature and pressures. As we have seen, the force of gravity at its surface is 27·65 times that at the surface of the Earth, where a body falls 16·1 feet in the first second of time; on the Sun, therefore, a body would fall 445 feet in the first second; and the atmospheric motions generally would be accelerated in the same proportion.

The high temperatures, the great pressures, the violent commotions which prevail on the Sun are, therefore, the direct consequence of its enormous mass. The Sun is, then, not merely the type and example of the chief source of light and heat in a given planetary system; it indicates to us that size and mass are the primary tokens by which we may judge the temperature of a world, and the activity to be expected in its changes.

CHAPTER IV

THE DISTRIBUTION OF THE ELEMENTS IN SPACE

It is now an old story, but still possessing its interest, how Fraunhofer analysed the light of the Sun by making it pass through a narrow slit and a prism, and found that the broad rainbow-tinted band of light so obtained was interrupted by hundreds of narrow dark lines, images in negative of the slit; and how Kirchhoff succeeded in proving that two of these dark lines were caused by the white light of the solar photosphere having suffered absorption at the Sun by passing through a stratum of glowing sodium vapour. From that time forward it has been known that the Sun is surrounded by an atmosphere of intensely heated gases, among which figure many of those elements familiar to us in the solid form on the Earth, such as iron, cobalt, nickel, copper, manganese, and the like. These metals, here the very types of solid bodies, are permanent gases on the Sun.

The Sun, then, is in an essentially gaseous condition, enclosed by the luminous shell which we term the photosphere. This shell Prof. C. A. Young and the majority of astronomers regard as consisting of a relatively thin layer of glowing clouds, justifying the quaint conceit of R. A. Proctor, who spoke of the Sun as a "Bubble"; that is, a globe of gas surrounded by an envelope so thin in comparison as to be a mere film. There has been much difference of opinion as to the substance forming these clouds, but the theory is still widely held which was first put forward by Dr. Johnstone Stoney in 1867, that they are due to the condensation of carbon, the most refractory of all known elements. Prof. Abbot, however, refuses to believe in a surface of this nature, holding that the temperature of the Sun is too high even at the surface to permit any such condensation.

The application of the spectroscope to astronomy is not confined to the Sun, but reaches much further. The stars also yield their spectra, and we are compelled to recognize that they also are suns; intensely heated globes of glowing gas, rich in the same elements as those familiar to us on the Earth and known by their spectral lines to be present on the Sun. The stars, therefore, cannot themselves be inhabited worlds any more than the Sun, and at a stroke the whole of the celestial luminaries within the furthest range of our most powerful telescopes are removed from our present search. Only those members of our solar system that shine by reflecting the light of the Sun can be cool enough for habitation; the true stars cannot be inhabited, for, whatever their quality and order, they are all suns, and must necessarily be in far too highly heated a condition to be the abode of life. Many of them may, perhaps, be a source of light and heat to attendant planets, but there is no single instance in which such a planet has been directly observed; no dark, non-luminous body has ever been actually seen in attendance on a star. Many double or multiple stars are known, but these are all instances in which one sun-like body is revolving round another of the same order. We see no body shining by reflected light outside the limits of the solar system. Planets to the various stars may exist in countless numbers, but they are invisible to us, and we cannot discuss conditions where everything is unknown. Enquiry in such a case is useless, and speculation vain.

The stars, as revealed to us by the spectroscope are all of the same order as the Sun, but they are not all of the same species. Quite a large number of stars, of which Arcturus is one of the best-known examples, show spectra that are essentially the same as that of the Sun, but there are other stars of which the spectra bear little or no semblance to[Pg 36] it. Nevertheless, it remains true that, on the whole, stellar spectra bear witness to the presence of just the same elements as we recognize in the Sun, though not always in the same proportions or in the same conditions—hydrogen, calcium, sodium, magnesium, iron, titanium, and many more are recognized in nearly all. It is true that not all the known terrestrial elements have yet been identified in either Sun or stars; but, in general, those missing are either "negative" elements like the halogens, or elements of great atomic weight like mercury and platinum. That elements of one class should, as a rule, reveal their presence in Sun and stars wherever these are placed, and, correspondingly, that other classes should as generally fail to show themselves, indicate that such absence is more likely to be due to the general structure of the stellar photospheres and reversing layers than to any irregularity in the distribution of matter in the universe. It is easy, for example, to conceive that the heavy metals may lie somewhat deeper down within the Sun or star than those of low atomic weight. In the case of the Sun, there seems a clear connection between atomic weight and the distinctness with which the element is recognized in the spectrum of the photosphere, the lower atomic weights showing themselves more conspicuously.

It is clear that not all elements present in a Sun or star show themselves in its spectrum. Oxygen is very feebly represented by its elemental lines, but the flutings of titanium oxide are found in sunspots, and with great distinctness in a certain type of stars. Nitrogen, too, though not directly recognized, proves its presence by the lines of cyanogen. The case of helium is one of particular interest; this element was recognized by a very bright yellow line in the solar prominences before it was known to exist on the Earth; indeed, it received the name *helium* because it then seemed to be a purely solar constituent. Now it is seen as a strong absorption line in the spectrum of many stars; but for some reason it is not in general seen as an absorption line over the Sun's disc, and if our Sun were removed to such distance so as to appear to us only as a star, we should have no evidence that it contained any helium at all. So far, then, as the evidence of the spectroscope goes, the elements present in the Earth are present throughout the whole extent of the universe within our view: the same elements and with the same qualities. For the lines of the spectrum of an element are the revelation of its innermost molecular structure, so that we can confidently affirm that hydrogen and oxygen on Sirius, Arcturus, or the Sun, are essentially the same elements as hydrogen and oxygen on the Earth. On a planet attached to any of these stars, the two gases would combine together to form water under just the same conditions as they do here on the Earth; and at suitable [Pg 38]temperatures that water would be a neutral liquid, capable of dissolving just the same chemical substances that it does here. It would freeze as it does here; it would evaporate as it does here; it would be water as completely in all its qualities and conditions as earthly water is. And what applies to one element or compound applies to all. Throughout the whole extent of space, the same building materials have been employed, and throughout they retain the same qualities.

Hydrogen is seen in the spectra of nearly all stars, and also in those of nebulæ. The elemental lines of oxygen are not indeed seen in stellar spectra, but that the element is present is shown by the flutings of titanium oxide which distinguish stars like Antares.

Nitrogen and carbon again are not recognized by their elemental lines, but the lines of cyanogen are seen in the spectra of comets and of sunspots, and hydrocarbon flutings in the spectra of comets and red stars; while in a few of the hottest stars even sulphur has recently been identified. All the five organo-genetic elements are therefore abundantly diffused through space; the materials for protoplasm, "the albuminous substance with water," are at hand everywhere. This being so, it is reasonable to infer that if organic life exists elsewhere than on this Earth, its essential feature, there as here, is the metabolism of nitrogenous carbon compounds in association with protoplasm.

But it is objected that "we are not yet able to identify all the lines in solar or stellar spectra; may not some of these lines be due to elements of which we know nothing here, and may not such new elements form complex and unstable compounds with each other, or with some of those familiar to us, that would take the place of the five organo-generators, and so give rise to a physical basis of life, different from that we know on this Earth?"

But the development of Mendeléeff's Periodic Law has shown that the elements are not to be regarded as disconnected entities. The Law as given in Mendeléeff's own words, runs: "The properties of the elements as well as the forms and properties of their compounds are in periodic dependence on, or (expressing ourselves algebraically) form a periodic function of the atomic weights of the elements." In other words, they form a series, not only as it regards their atomic weights, but also as it regards their own properties and the forms and properties of their compounds. We are no longer at liberty, as we might have been many years ago, to call into fancied existence new elements having no relation in their properties and compounds to those with which we are acquainted. New elements, no doubt, will be discovered in the future, as in the past; and indeed we may be able to discover them and learn their atomic weights and properties without ever being able to handle them in a terrestrial laboratory.

In a series of remarkable papers communicated to the Royal Astronomical Society during the past year (1911-1912), Dr. J. W. Nicholson has given the result of his computation of the positions of the spectral lines of two elements of simple structure, and has found that the resulting lines correspond, for one dynamical system, to the chief unidentified lines observed in the spectra of nebulæ, and for the other, to the chief unidentified lines in the spectrum of the corona. The latter element is probably associated with the halogens, but of much lower atomic weight (namely, 1·3), than fluorine; he therefore gives it the name of *Protofluorine*. The other element, to which he gives the name *Nebulium*, will have an atomic weight of 2·1. Prof. Max Wolf, of Heidelberg, has recently pointed out the evidence of the presence of two other unknown gases in the Ring nebula in Lyra, and there is no reason to suppose that the process of discovery has come to an end. But we cannot imagine that we shall discover any new elements that are more abundant and more universally diffused than the five which give us protoplasm—"the physical basis of life." To take an analogy from the solar system: many hundreds of planetoids have now been discovered between the orbits of Mars and Jupiter, and probably many hundreds more remain to be discovered; but of one thing we are certain, that none of the planetoids yet to be discovered will be of the same rank as either of those two guardians, Mars and Jupiter, who revolve on the confines of the planetoidal zone. Indeed, Ceres, the planetoid first discovered, has a greater mass than the aggregate of all discovered since, and probably of all that exist in the zone.

Water is essential for life here, but the quality in water which restricts the range of terrestrial life is that it freezes at 0° Centigrade, and boils at 100° Centigrade; it is only in the liquid state during the intermediate range of 100 degrees. In order to extend the range for living organisms, we should have, therefore, to discover a new vehicle, that, possessing all the other qualities of water, is not restricted to the liquid state within the same limits. But we are at once met with the difficulty that the first essential for the vehicle is that it should be abundant, and there are no other elements more abundant than hydrogen and oxygen. This new vehicle must, like water, be both neutral and stable, or it would itself interfere with the highly unstable compounds that are a necessity for metabolism. And, if we could find this new vehicle, liquid at temperatures outside the 0° to 100° Centigrade, have we any reason to suppose that protoplasm itself would be able to endure these outlying temperatures? Looking through the range of substances available, we can only say that none other presents itself as approaching water in suitability for its essential office. If we, ourselves, were able to create a vehicle, could we imagine one more perfectly suited?

CHAPTER V

THE MOON

The Sun and Moon offer to our sight almost exactly the same apparent diameters; to the eye, they look the same size. But as we know the Sun to be 400 times as distant as the Moon, it is necessarily 400 times as large; its surface must exceed that of the Moon by the square of 400, or 160,000; its volume by the cube of 400, or 64,000,000. As the Sun is of low mean density, its mass does not exceed that of the Moon in quite the same high ratio; but it is equal in mass to

27,000,000 moons.

Compared with the Sun, the Moon is therefore an insignificant little ball—a mere particle; but as a world for habitation it possesses some advantages over the Sun. The first glance at it in a telescope is sufficient to assure the observer that he is looking at a solid, substantial globe. It is not only substantial, it is rugged; its surface is broken up into mountains, hills, valleys, and plains; the mountains stand out in sensible relief; it looks like a ball of solid silver boldly embossed and chased.

So far all is to the good for the purpose of habitation. Wherever men are, they must have a solid platform on which to stand; they must have a stable terrene whereon their food may grow, and this the Moon could supply. "The Earth's gloom of iron substance" is necessary for man here, and the Moon appears to offer a like stability.

Another favourable condition is that we know that the Moon receives from the Sun a sufficient supply of light and heat. Each square yard of its surface receives, on the average, the same amount of light and heat that would fall upon a square yard on the Earth that was presented towards the Sun at the same inclination; and we know from our own experience that this is sufficient for the maintenance of life.

And the Moon is near enough for us to subject her to a searching scrutiny. Every part of the hemisphere turned toward us has been repeatedly examined, measured, and photographed; to that extent our knowledge of its topography is more complete than of the world on which we live. There are no unexplored regions on our side of the Moon. The great photographs taken in recent years at the observatories of Paris and of the University of Chicago have shown thousands of "crater-pits," not more than a mile across; and narrow lines on the Moon's surface have been detected with a breadth less than one-tenth of this. An elevation on the Moon, if it rose up abruptly from an open plain, would make its presence apparent by the shadow which it would cast soon after sunrise or near sunset; in this way an isolated building, if it were as large as the great pyramid of Ghizeh, would also show itself, and all our great towns and cities would be apparent as areas of indistinct mottling, though the details of the cities would not be made out.

But if vegetation took the same forms on the Moon as on the Earth, and passed through the same changes, we should have no difficulty in perceiving the evidence of its presence. If we were transported to the Moon and turned our eyes earthward, we should not need the assistance of any telescope in order to detect terrestrial changes which would be plainly connected with the seasonal changes of vegetation. The Earth would present to us a disc four times the apparent diameter of the Moon, and on that disc Canada would offer as great an area as the whole of the Moon does to us. We could easily follow with the naked eye the change from the glittering whiteness of the aspect of Canada when snow-covered in winter, to the brown, green and gold which would succeed each other during the brighter months of the year. And this type of change would alternate between the northern and southern hemispheres, for the winter of Canada is the summer of the Argentine, and conversely.

We ought, therefore, to have no difficulty in observing seasonal changes on the Moon, if such take place. But nothing of the kind has ever been remarked; no changes sufficiently pronounced for us to be sure of them are ever witnessed. Here and there some slight mutations have been suspected, nearly all accomplishing their cycle in the course of a lunar day; so that it is difficult to separate them from changes purely apparent, brought about by the change in the incidence of the illumination.

The difference in appearance of a given area on the Moon when viewed under a low Sun and when the Sun is on the meridian is very striking. In the first case everything is in the boldest relief; the shadows are long and intensely black; the whole area under examination in the telescope seems as if it might be handled. Under the high Sun, the contrasts are gone; the scenery appears flat, many of the large conspicuous markings are only recognized with difficulty. Thus the terse remark of Mädler, "The full Moon knows no Maginus," has become a proverb amongst selenographers; yet Maginus is a fine walled plain some eighty miles in diameter, and its rampart attains a height in parts of 14,000 feet. Maginus lies near Tycho, which has been well named "the lunar metropolis," for from it radiates the principal system of bright streaks conspicuous on the full Moon. These white streaks appear when the shadows have vanished or are growing short; they are not seen under a low Sun.

The changes which appear to take place in the lunar formations owing to the change in their illumination are much more striking and varied than would be anticipated. But the question arises whether all the changes that are associated with the progress of the lunar day can be ascribed to this effect. Thus, Prof. W. H. Pickering writes concerning a well-known pair of little craters of about nine miles in diameter, "known as Messier and Messier A, situated side by side not far from the centre of the Mare Fecunditatis. When the Sun rises first on them, the eastern one, A, is triangular and larger than Messier, which latter is somewhat pear-shaped. About three days after sunrise they both suddenly turn white, Messier rapidly grows in size, soon surpasses A, and also becomes triangular in shape. Six days after sunrise the craters are again nearly of the same size, owing to the diminution of Messier. The shape of A has become irregular, and differs in different lunations. At nine days after sunrise the craters are exactly alike in size and shape, both now being elliptical, with their major axes lying in a nearly N. and S. direction. Just before sunset A is again the larger, being almost twice the size of Messier."

Some observers explain this cycle of changes as due merely to the peculiar contour of the two[Pg 48] objects, the change in the lighting during the lunar day altering their apparent figures. Prof. W. H. Pickering, on the other hand, while recognizing that some portion of the change of shape is probably due to the contour of the ground, conceives that, in order to explain the whole phenomenon, it is necessary to suppose that a white layer of hoar frost is formed periodically round the two craters. It is also alleged that whereas Mädler described the two craters as being exactly alike eighty years ago, Messier A is now distinctly the larger; but it is very doubtful whether Mädler's description can be trusted to this degree of nicety. If it could, this would establish a permanent change in the actual structure of the lunar surface at this point.

There are several other cases of the same order of ambiguity. The most celebrated is Linné, a white spot about six miles in diameter on the Mare Serentatis. This object appears to change in size during the progress of the lunar day, and, as with Messier, some selenographers consider that it has also suffered an actual permanent change in shape within the last sixty or seventy years. Here again the evidence is not decisive; Neison is by no means convinced that a change has taken place, yet does not think it impossible that Linné may once have been a crater with steep walls which have collapsed into its interior through the force of gravity.

Another type of suspected change is associated with the neighbourhood of Aristarchus, the brightest formation on the Moon, so bright indeed that Sir William Herschel, observing it when illuminated by earthshine in the dark portion of the Moon, thought that he was watching a lunar volcano in eruption. In 1897, on September 21, the late Major Molesworth noticed that the crater was at that time under the rays of the setting Sun, and filled with shadow, and the inner terraces, which should have been invisible, were seen as faint, knotted, glimmering streaks under both the eastern and western walls, and the central peak was also dimly discernible. He thought this unusual lighting up of rocks on which the Sun had already set might be due either to phosphorescence produced by long exposure to the Sun's rays, or to inherent heat, or to reflected glare from the western rampart. Still more important, both Major Molesworth and Mr. Walter Goodacre, each on more than one occasion, observed what seemed to be a faint bluish mist on the inner slope of the east wall, soon after sunrise, but this was visible only for a short time. Other selenographers too, on rare occasions, have made observations accordant with these, relating to various regions on the Moon.

These, and a few other similar instances, are all that selenography has to offer by way of evidence of actual lunar change. Of seeming change there is abundance, but beyond that we have only cases for controversy, and one of the most industrious of the present-day observers of the Moon, M.Philip Fauth, declares that "as a student of the Moon for the last twenty years, and as probably one of the few living investigators who have kept in practical touch with the results of selenography, he is bound to express his conviction that no eye has ever seen a physical change in the plastic features of the Moon's surface."

In this matter of change, then, the Earth and Moon stand in the greatest contrast to each other. As we have seen, from the view-point of the Moon, the appearance of the Earth would change so manifestly with the progress of the seasons that no one could fail to remark the difference, even though observing with the naked eye. But from the view-point of the Earth, the Moon when examined by our most experienced observers, armed with our most powerful telescopes, offers us only a few doubtful enigmatical instances of possible change confined to small isolated localities; we see no evidence that the "gloom of iron substance" below is ever concealed by a veil of changing vegetation, or that "between the burning light and deep vacuity" of the heavens above, the veil of the flying vapour has ever been spread out. We see the Moon so clearly that we are assured it holds no water to nourish plant life; we see it so clearly because there is no air to carry the vapour that might dim our view.

Life is change, and a planet where there is no change, or where that change is very small, can be no home for life. The "stability and insensibility" are indeed required in the platform upon which life is to appear, but there must be the presence of "the passion and the perishing," or life will be unable to find a home.

We infer the absence of water and air from the Moon not only from the unchanging character of its features and the distinctness with which we see them; we are able to make direct observations. Galileo, the first man to observe the Moon to better advantage than with the naked eye, was not long before he decided that the Moon contained no water, for though Milton, in a well-known passage, makes Galileo discover

"Rivers or mountains on her spotty globe,"

Galileo himself wrote: "I do not believe that the body of the Moon is composed of earth and water." The name of *maria* was given to the great grey plains of the Moon by Hevelius, but this was simply for convenience of nomenclature, not because he actually believed them to be seas. One observation is, in itself, sufficient to prove that the maria are not water surfaces. The Moon's "terminator," that is to say, the line dividing the part in sunlight from that in darkness, is clearly irregular when it passes over the great plains; were they actually sea it would be a bright line and perfectly smooth. The grey plains are therefore not expanses of water now, nor were they in time past. It is obvious that in some remote antiquity their surface was in a fluid condition, but it was the fluidity of molten rock. This is seen by the way in which the maria have invaded, breached, broken down, and submerged many of the circular formations on their margins. Thus the Mare Humorum has swept away half the wall of the rings, Hippalus and Doppelmayer, and far out in the open plain of the Mare Nubium, great circles like Kies, and that immediately north of Flamsteed, stand up in faint relief as of half-submerged rings. Clearly there was a period after the age in which the great ring mountains and walled plains came into existence, when an invasive flood attacked and partially destroyed a large proportion of them. And the flood itself evidently became more viscous and less fluid the further it spread from its original centre of action, for the ridges and crumpling of the surface indicate that the material found more and more difficulty in its flow.

We have evidence just as direct that there is no atmosphere. This is very strikingly shown when the Moon, in its monthly progress among the stars, passes before one of them and occults it. Such an occultation is instantaneous, and is particularly impressive when either a disappearance or a reappearance occurs at the defective limb; that is to say, at the limb which is not illuminated by the Sun, and is therefore invisible. The observer may have a bright star in the field of view, showing steadily in a cloudless sky; there is not a hint of a weakening in its light; suddenly it is gone. The first experience of such an observation is most disconcerting; it is hardly less disconcerting to observe the reappearance at the dark limb. One moment the field of view of the telescope is empty; the next, without any sort of dawning, a bright star is shining steadily in the void, and it almost seems to the observer as if an explosion had taken place. If the Moon had an atmosphere extending upwards from its surface in all directions and of any appreciable density, an occultation would not be so exceedingly abrupt; and, in particular, if the occultation were watched through a spectroscope, then, at the disappearance, the spectrum of the star would not vanish as a whole, but the red end would go first, and the rest of the spectrum would be swept out of sight successively, from orange to the violet. This does not happen; the whole spectrum goes out together, and it is clear that no appreciable atmosphere can exist on the Moon. In actual observation so inappreciable is it that its density at the Moon's surface is variously estimated as 1⁄300th of that of the Earth by Neison, and as 1⁄10000th by W. H. Pickering. If the Moon possessed an atmosphere bearing the same proportion to her total mass as we find in the case of the Earth, she would have a density of one-fortieth of our atmosphere at the sea level.

The Moon is at the same mean distance from the[Pg 54] Sun as the Earth, and therefore, surface for surface, receives from it on the average the same amount of light and heat. But it makes a very different use of these supplies. Bright as the Moon appears when seen at the full on some winter night, it has really but a very low power of reflection, and is only bright by contrast with the darkness of the midnight sky. If the full Moon is seen in broad daylight, it is pale and ghost-like. Sir John Herschel has put it on record that when in South Africa he often had the opportunity of comparing the Moon with the face of Table Mountain, the Sun shining full upon both, and the Moon appeared no brighter than the weathered rock. The best determinations of the *albedo* of the Moon, that is to say, of its reflective power, give it as 0·17, so that only one-sixth of the incident light is reflected, the other five-sixths being absorbed. It is difficult to obtain a good determination of the Earth's *albedo*, but the most probable estimate puts it as about 0·50, or three times as great as that of the Moon. This high reflective power is partly to be accounted for by the great extent of the terrestrial polar caps, but chiefly by the clouds and dust layer always present in its atmosphere.

A larger proportion, therefore, of the solar rays are employed in heating the soil of the Moon than in heating that of the Earth, and in this connection the effect of an important difference between the two worlds must be noted. The Earth rotates on its axis in 23 hours 56 minutes 4 seconds, the mean length of its rotation as referred to the Sun being 24 hours. The rotation of the Moon, on the other hand, takes 27 days 7 hours 43 minutes to accomplish, giving a mean rotation, as referred to the Sun, of 29 days 12 hours 44 minutes. The lunar surface is therefore exposed uninterruptedly to the solar scorching for very nearly fifteen of our days at a time, and it is, in turn, exposed to the intense cold of outer space for an equal period. As the surface absorbs heat so readily, it must radiate it as quickly; hence radiation must go on with great rapidity during the long lunar night. Lord Rosse and Prof. Very have both obtained measures of the change in the lunar heat radiation during the progress of a total eclipse of the Moon, with the result that the heat disappeared almost completely, though not quite at the same time as the light. Prof. Langley succeeded in obtaining from the Moon, far down in the long wave lengths of the infra-red, a heat spectrum which was only partly due to reflection from the Sun; part coming from the lunar soil itself, which, having absorbed heat from the Sun, radiated it out again almost immediately. In 1898, Prof. Very, following up Langley's line of work, concluded that the temperature of the lunar soil must range through about 350° Centigrade, considerably exceeding 100° at the height of the lunar day, and falling to about the temperature of liquid air during the lunar night. So wide a range of temperature must be fatal to living organisms, particularly when the range is repeated at short, regular intervals of time. But this range of temperature comes directly from the length of the Moon's rotation period; for the longer the day of the Moon, the higher the temperature which may be attained in it; the longer the night, the greater the cold which will in turn be experienced. We learn, therefore, that the time of rotation of a planet is an important factor in its habitability.

CHAPTER VI

THE CANALS OF MARS

Both of the two worlds best placed for our study are thus, for different reasons, ruled out of court as worlds for habitation. The Sun by its vastness, its intolerable heat and the violence of its changes, has to be rejected on the one hand, while the Moon, so small, and therefore so rigid, unchanging and bare, is rejected on the other.

Of the other heavenly bodies, the planet Mars is the one that we see to best advantage. Two other planets, Eros and Venus, at times come nearer to us, but neither offers us on such occasions equal facilities for their examination. But of Mars it has been asserted not only that it is inhabited, but that we know it to be the case, since the evidence of the handiwork of intelligent beings is manifest to us, even across the tremendous gulf of forty or more million miles of space.

A claim so remarkable almost captures the position by its audacity. There is a natural desire among men to believe the marvellous, and the very boldness of the assertion goes no small way to overcome incredulity. And when we consider how puny are men as we see them on this our planet, how minute their greatest works, how superhuman any undertaking would be which could demonstrate our existence to observers on another planet, we must admit that it is a marvel that there should be any evidence forthcoming that could bear one way or another on the solution of a problem so difficult.

The first fact that we have to remember with regard to the planet Mars is the smallness of its apparent size. To the eye it is nearly a star—a point of light without visible surface. It is almost twice the size of the Moon in actual diameter, but as its mean distance from the Earth is 600 times that of the Moon, its mean apparent diameter is 300 times smaller. We cannot, however, watch Mars in all parts of its orbit; it is best placed for observation, and, therefore, most observed, when in opposition, and oppositions may be favourable or unfavourable. At the most favourable opposition, Mars is 140 times as distant as the Moon; at the least favourable, 260 times; so that on such occasions its apparent size varies from 1/70th of the diameter of the Moon to 1/130th. But a telescope with a magnifying power of 70 could never, under the most perfect conditions, show Mars, even in the closest opposition, as well as the Moon is seen with the naked eye, for the practical magnifying power of a telescope is never as great as the theoretical. In practice, a child's spy-glass magnifying some six diameters will show the full Moon to better advantage than Mars has ever been seen, even in our most powerful telescopes.

The small apparent size of the planet explains how it was that Galileo does not seem to have been able to detect any markings upon it. In 1659, Huyghens laid the foundation stone of areography by observing some dark spots, and determining from their apparent movements that the planet had a rotation on its axis, which it accomplished in about the same time as the Earth. Small and rough as are the drawings that Huyghens made, the identification of one or two of his spots is unmistakable. Seven years later, in 1666, both Cassini and Hooke made a number of sketches, and those by Hooke have been repeatedly used in modern determinations of the rotation period of the planet. The next great advance was made by Sir William Herschel, who, during the oppositions of 1777, 1779, 1781, and 1783, determined the inclination of the axis of Mars to the plane of its orbit, measured its polar and equatorial diameters, and ascertained the amount of the polar flattening. He paid also special attention to two bright white spots upon the planet, and he showed that these formed round the planet's poles and increased in size as the winter of each several hemisphere drew on and diminished again with the advance of summer, behaving therefore as do the snow caps of our own polar regions.

The next stage in the development of our knowledge of Mars must be ascribed to the two German astronomers, Beer and Mädler, who made a series of drawings in the years 1830, 1832 and 1837, by means of a telescope of 4 inches aperture, from which they were able to construct a chart of the entire globe. This chart may be considered classic, for the features which it represents have been observed afresh at each succeeding opposition. Mars, therefore, possesses a permanent topography, and some of the markings in question can be identified, not only in the rough sketches made by Sir William Herschel, but even in those made by Hooke and Cassini as far back as the year 1666. In the forty years that followed, the planet was studied by many of the most skilled observers, particularly by Mr. J. N. Lockyer in 1862, and the Rev. W. R. Dawes in 1864. In 1877, the late Mr. N. E. Green, drawing-master to Queen Victoria, and a distinguished painter in water colours, made a series of sketches of the planet from a station in the island of Madeira 2000 feet above sea-level. When the opposition was over, Mr. Green collected together a large number of drawings, and formed a chart of the planet, much richer in detail than any that had preceded it, and from his skill, experience and training as an artist he reproduced the appearance of the planet with a fidelity that had never been equalled before and has never been surpassed since. At this time it was generally assumed that Mars was a miniature of our own world. The brighter districts of its surface were supposed to be continents, the darker, seas. As Sir William Herschel had already pointed out long before, the little world evidently had its seasons, its axis being inclined to the plane of its orbit at much the same angle as is the case with the Earth; it had its polar caps, presumably of ice and snow; its day was but very little longer than that of the Earth; and the only important difference seemed to be that it had a longer year, and was a little further off the Sun. But the general conclusion was that it was so like the Earth in its conditions that we had practically found out all that there was to know; all that seemed to be reserved for future research was that a few minor details of the surface might be filled in as the power of our telescopes was increased.

But fortunately for progress, this sense of satisfaction was to be rudely disturbed. As Mars, in its progress round the Sun, receded from the Earth, or rather as the Earth moved away from it, the astronomers who observed so diligently during the autumn of 1877 turned their attention to other objects. One of them, however, Schiaparelli, the most distinguished astronomer on the continent[of Europe, still continued to watch the planet, and, as the result of his labours, he published some months later the first of a magnificent series of *Memoirs,* bringing to light what appeared to be a new feature. His drawings not only showed the "lands" and "seas," that is to say the bright and dark areas, that Green and his predecessors had drawn, but also a number of fine, narrow, dark lines crossing the "lands" in every direction. These narrow lines are the markings which have since been so celebrated as the "canals of Mars," and the discussion as to the real nature of these canals has focussed attention upon Mars in a way that, perhaps, nothing else could have done. Before 1877 the study of planetary markings was left almost entirely to the desultory labours of amateurs, skilled though many of them were; since 1877, the most powerful telescopes of the great public observatories of the world have been turned upon Mars, and the most skilful and experienced of professional astronomers have not been ashamed to devote their time to it.

There is no need to pass in review the whole of the immense mass of observations that have been accumulated since Schiaparelli brought out the first of his great Memoirs. That Memoir gave rise to an immediate controversy, for many astronomers of skill and experience had observed the planet in 1877 without detecting the network of lines which Schiaparelli had revealed, and it was natural that they should feel some reluctance in accepting results so strange and novel. But little by little this controversy has passed. We now know that the "canals" vary much in their visibility, and "curiously enough the canals are most conspicuous, not at the time the planet is nearest to the Earth and its general features are in consequence best seen, but as the planet goes away the canals come out. The fact is that the orbital position and the seasonal epoch conspire to a masking of the phenomena." This was the chief reason why Schiaparelli's discoveries seemed at first to stand so entirely without corroboration; the "canals" did not become conspicuous until after most observers had desisted from following the planet. Another reason was that, in 1877, Mars was low down in the sky for northern observatories, and good definition is an essential for their recognition. But the careful examination of drawings made in earlier oppositions, especially those made by Dawes and Green, afforded confirmation of not a few of Schiaparelli's "canals"; even in 1877 a few of the easiest and most conspicuous had been delineated by other astronomers before any rumour of Schiaparelli's work had come abroad, and as Mars came under observation again and again at successive oppositions, the number of those who were able to verify Schiaparelli's discoveries increased. It has now long been knownthat the great Italian astronomer was not the victim of a mere optical illusion; there were actual markings on the planet Mars where he had represented them; markings which, when seen under like conditions and with equal instrumental equipment, did present the appearance of straight, narrow lines. The "canals of Mars" are not mere figments of the imagination, but have a real objective basis.

As this controversy has passed away, another and a very different one has arisen out of an unfortunate mistranslation of the term chosen by Schiaparelli to indicate these linear streaks. In conformity with the type of nomenclature adopted by previous areographers who had divided Mars into "seas," "continents," "islands," "isthmuses," "straits" and the like, Schiaparelli had called the narrow lines he detected "*canali*", that is to say "channels," but without intending to convey the idea of artificial construction. Indeed, he himself was careful to point out that these designations "were not intended to prejudge the nature of the spot, and were nothing but an artifice for helping the memory and for shortening descriptions." And he added, "We speak in the same way of the lunar seas, although we well know that there are no true seas on the Moon." But "*canali*" was unhappily rendered in English as "canals," instead of "channels." "Channel" would have left the nature of the marking an open question, but, in English, "canal" means an artificial waterway. Here then the question as to whether or no Mars is inhabited comes definitely before us. Have we sufficient grounds for believing that the "canals" are artificial constructions, or may they be merely natural formations?

In 1894, Mr. Percival Lowell founded at Flagstaff, Arizona, U.S.A., a well-equipped observatory for the special study of Mars, and he has continued his scrutiny of the planet from that time to the present with the most unrelaxing perseverance. The chief results that he has obtained have been the detection of many new "canals"; the discovery of a number of dark, round dots, termed by him "oases," at the junctions of the "canals"; and the demonstration that the "canals" and certain of the dusky regions are subject to strictly seasonal change, as really as the polar caps themselves. In addition, he has formed the conclusion, which he has supported with much ingenuity and skill, that the regularity of the "canals" and "oases" quite precludes the possibility of their being natural formations. Hence there has arisen the second controversy: that on the nature of the "canals"; for Mr. Lowell considers that their presence proves the existence of inhabitants on Mars, who, by means of a Titanic system of irrigation, are fighting a losing battle against the gradual desiccation of their planet.

"Organic life needs water for its existence. This water we see exists on Mars, but in very scant amount, so that if life of any sort exists there, it must be chiefly dependent on the semi-annual unlocking of the polar snows for its supply, inasmuch as there are no surface bodies of it over the rest of the planet. Now the last few years, beginning with Schiaparelli in 1877, and much extended since at Flagstaff, have shown:

"The surface of the planet to be very curiously meshed by a fine network of lines and spots.

"Now if one considers first the appearance of this network of lines and spots, and then its regular behaviour, he will note that its geometrism precludes its causation on such a scale by any natural process and, on the other hand, that such is precisely the aspect which an artificial irrigating system, dependent upon the melting of the polar snows, would assume. Since water is only to be had at the time it is there unlocked, and since for any organic life it must be got, it would be by tapping the disintegrated cap, and only so, that it could be obtained. If Mars be inhabited, therefore, it is precisely such a curious system we should expect to see, and only by such explanation does it seem possible to account for the facts.

"These lines are the so-called canals of Mars. It is not supposed that what we see is the conduit itself. On the contrary, the behaviour of these lines indicates that what we are looking at is vegetation. Now, vegetation can only be induced by a water-supply. What we see resembles the yearly inundation of the Nile, of which to a spectator in space the river itself might be too narrow to be seen, and only the verdured country on its banks be visible. This is what we suppose to be the case with Mars. However the water be conducted, whether in covered conduits, which seems probable, or not, science is not able to state, but the effects of it are so palpable and so exactly in accord with what such a system of irrigation would show, that we are compelled to believe that such is indeed its *vera causa*."

Beside the bulky *Memoirs* in which Prof. Lowell has published the scientific results obtained at his observatory at Flagstaff, and papers and articles appearing in various scientific journals, he has brought out three books of a more popular character: *"Mars"*; *"Mars and its Canals"*; and *"Mars as the Abode of Life."* In these he shows that to the assiduity of the astronomer he adds the missionary's zeal and eagerness for converts as he pleads most skilfully for the acceptance of his chosen doctrine of the presence of men on Mars. In the last of the three books mentioned, he deals directly with "Proofs of Life on Mars." The presence of vegetation may be inferred from seasonal changes of tint, just as an observer on the Moon might with the naked eye watch effects on the Earth. But though "vegetable life could thus reveal itself directly, animal life could not. Not by its body but by its mind would it be known. Across the gulf of space it could be recognized only by the imprint it had made on the face of Mars."

"Confronting the observer are lines and spots that but impress him the more, as his study goes on, with their non-natural look. So uncommonly regular are they, and on such a scale as to raise suspicions whether they can be by nature regularly produced" (p. 188).

"... Unnatural regularity, the observations showed, betrays itself in everything to do with the lines: in their surprising straightness, their amazing uniformity throughout, their exceeding tenuity, and their immense length"

"As a planet ages, its surface water grows scarce. Its oceans in time dry up, its rivers cease to flow, its lakes evaporate Now, in the struggle for existence, water must be got.... Its procuring depends on the intelligence of the organisms that stand in need of it.... As a planet ages, any organisms upon it will share in its development. They must evolve with it, indeed, or perish. At first they change only, as environment offers opportunity, in a lowly, unconscious way. But, as brain develops, they rise superior to such occasioning.... The last stage in the expression of life upon a planet's surface must be that just antecedent to its dying of thirst.... With an intelligent population this inevitable end would be long foreseen.... Both polar caps would be pressed into service in order to utilize the whole available supply and also to accommodate most easily the inhabitants of each hemisphere"

[Pg 69]"That intelligence should thus mutely communicate its existence to us across the far reaches of space, itself remaining hid, appeals to all that is highest and most far-reaching in man himself. More satisfactory than strange this; for in no other way could the habitation of the planet have been revealed. It simply shows again the supremacy of mind.... Thus, not only do the observations we have scanned lead us to the conclusion that Mars at this moment is inhabited, but they land us at the further one that these denizens are of an order whose acquaintance was worth the making" For the moment, let us leave Prof. Lowell's argument as he puts it. Whether we accept it or not, it remains that it is a marvellous achievement of the optician's skill and the observer's devotion that from a planet so small and so distant as Mars any evidence should be forthcoming at all that could bear upon the question of the existence of intelligent organisms upon its surface. But it is of the utmost significance to note that the whole question turns upon the presence of water—of water in the liquid state, of water in a sufficient quantity; and the final decision, for Mr. Lowell's contention, or against it, must turn on that one point. The search for Life on Mars is essentially a search for Water; a search for water, not only in the present state of Mars, but in its past as well. For, without water in sufficient quantities in the past, life on Mars could not have passed through[Pg 70] the evolutionary development necessary to its attaining its highest expression,—that where the material living organism has become the tabernacle and instrument of the conscious intelligent spirit.

CHAPTER VII

THE CONDITION OF MARS

The planet Mars is the debatable ground between two opinions. Here, the two opposing views join issue; the controversy comes to a focus. The point in debate is whether certain markings—some linear, some circular—are natural or artificial. If, it is argued, some are truly like a line, without curve or break, as if drawn with pen, ink, and ruler; or others, so truly circular, without deviation or break, as if drawn with pen, ink, and compass; if, moreover, when we obtain more powerful telescopes, erected in better climates for observing, these markings become more truly lines and circles the better we see them; then they are *artificial*, not natural structures.

But artificial structures imply artificers. And if the structures are so designed as to meet the needs of a living organism, it implies that the living organism that designed them must have a reasonable mind lodged in a natural body. If, then, the "lines" and "circles" that Prof. Lowell and his disciples assert to be artificial canals and oases are really such, they premise the order of being that we call Man. But these canals and oases also premise the liquid that we call Water—water that flows and water utilized in cultivation. In this chapter we will leave out of count the first premiss—Man—and only deal with what concerns the second premiss—Water; with water that flows and is utilized in vegetation.

PLANETARY STATISTICS

	Minor Planets.	Inner Planets.					Outer Planets.			
	Ceres	Moon	Mercury	Mars	Venus	Earth	Uranus	Neptune	Saturn	Jupiter
Proportions of the Planets:—										
Diameter in miles	477	2163	3030	4230	7700	7918	31900	34800	73000	86500
" ⊕ = 1	0·06	0·273	0·383	0·534	0·97	1·000	4·029	4·395	9·21	10·9

					2				9	24
Surface, ⊕ = 1	0·004	0·075	0·147	0·285	0·945	1·000	16·2	19·3	85·0	119·3
Volume, ⊕ = 1	0·0002	0·02	0·06	0·15	0·92	1·00	65·	85·	760·	1304·
Density, Water = 1	2·8 ?	3·39	4·72	3·92	4·94	5·55	1·22	1·11	0·72	1·32
" ⊕ = 1	0·5 ?	0·61	0·85	0·71	0·89	1·00	0·22	0·20	0·13	0·24
Mass, ⊕ = 1	0·0001	0·012	0·048	0·107	0·820	1·000	14·6	17·0	94·8	317·7
Gravity at surface, ⊕ = 1	0·028	0·17	0·33	0·38	0·87	1·00	0·90	0·89	1·18	2·65
Rate of Fall, Feet in the First Second	0·45	2·73	5·30	6·11	13·99	16·08	14·47	14·31	18·97	42·61
Albedo	0·14	0·17	0·14	0·22	0·76	0·50 ?	0·60	0·52	0·72	0·62
Details of Orbit:—										
Mean Distance from Sun in millions of miles	257·1	92·9	36·0	141·5	67·2	92·9	1781·9	2791·6	886·0	483·3
" " Earth's distance = 1	2·767	1·000	0·387	1·524	0·723	1·000	19·183	30·055	9·539	5·203
Period of Revolution, in years	4·60	1·00	0·24	1·88	0·62	1·00	84·02	164·78	29·46	11·86
Velocity, in miles per second	11·1	18·5	9·7	15·0	21·9	18·5	4·2	3·4	6·0	8·1
Eccentricity	0·0763	0·0168	0·2056	0·0933	0·0068	0·0168	0·0463	0·0090	0·0561	0·0483
Aphelion Distance, Perihelion = 1	1·157	1·034	1·517	1·207	1·013	1·034	1·097	1·018	1·107	1·101

Inclination of Equator(?) to Orbit		1° 32´	(?)	24° 0´	(?)	23° 27´	(?)	(?)	26° 49´	3° 5´
		d h m	d	h m s		h m s	h m		h m	h m
Rotation period	(?)	27 7 43	88(?)	24 37 23	(?)	23 56 9·4	30(?)	(?)	10 14±	9 55 ±
Atmosphere, assuming the total mass of the atmosphere to be proportional to the mass of the planet:—										
Pressure at the surface in lb. per sq. inch.	0·014	0·40	1·6	2·1	11·1	14·7	11·9	11·5	19·4	103·8
" " " in "atmospheres"	0·0009	0·027	0·108	0·143	0·754	1·000	0·81	0·78	1·32	7·06
Level of half surface pressure in miles	119·0	19·6	10·1	8·8	3·8	3·3	3·7	3·8	2·8	1·3
Boiling point of water at the surface		22°C	53°C	53°C	92°C	100°C	94°C	93°C	108°C	166°C
Temperature:—										
Light and heat received	0·13	1·00	6·67	0·43	1·91	1·00	0·003	0·001	0·011	0·037

from Sun, ⊕ = 1										
Reciprocal of square-root of distance, ⊕ = 1	0·60	1·00	1·61	0·81	1·18	1·00	0·23	0·18	0·32	0·44
Equatorial temp. of ideal planet, Absolute	188	312°	502°	253°	368°	312°	71°	56°	101°	137°
" " " " Centigrade	-65	+39	+229	-20	+95	+39	-202	-217	-172	-136
Average temp. of ideal planet, Absolute	174	290	467	235	342	290	66	52	94	127
" " " " Centigrade	-99	+17	+194	-38	+69	+17	-207	-221	-179	-146
Upper limit under zenith sun, Absolute	248	412	664	337	486	412	94	74	133	180
" " " " Centigrade	-25	+139	+391	+64	+213	+139	-179	-199	-140	-93
Average temp. of equivalent disc, Absolute	223	371	598	300	438	371	84	67	120	162
" " " " Centigrade	-50	+98	+325	+27	+165	+98	-189	-206	-153	-111

For in regard to this particular premiss we can do away with hypothesis, and deal only with certain physical facts that are not controversial and are not in dispute.

The first of this series of facts concerning Mars about which there can be no controversy or dispute relates to its size and mass. As the foregoing Table shows, it comes between the Moon and the Earth in these respects.

The figures show at a glance that Mars ranks in its dimensions between the Moon and the Earth, and that, on the whole, it is more like to the Moon than it is to the Earth.

But in what way would this affect Mars as a suitable home for life? In many ways; and amongst these the distribution of its atmosphere and the sluggishness of its atmospheric circulation are not the least important.

It was mentioned in Chapter III that at a height of about three and a third miles the barometer will stand at 15 inches, or half its mean height at sea level, showing that one half the atmosphere has been passed through. Mont Blanc, the highest mountain in Europe, is under 3 miles in height, so that it is not possible, in Europe, to climb to the level of half-pressure; Mt. Everest, the highest mountain in the world, is not quite six miles high, so that no part of the solid substance of our planet reaches up to the level of the quarter pressure. On a very few occasions daring aeronauts have soared into the empyrean higher than the summits of even our loftiest mountains, but the excursion has been a dangerous one, and they have with difficulty brought their life back from so rare and cold, so inhospitable a region. When Gay-Lussac, in 1804, attained a height of 23,000 feet above sea level, the thermometer, which on the ground read 31° C., sank to 9° below zero, and the rare atmosphere was so dry that paper crumpled up as if it had been placed near the fire, and his pulse rose to 120 pulsations a minute instead of his normal 66. When Mr. Glaisher and Mr. Coxwell made their celebrated ascent between 1 and 2 o'clock on the afternoon of September 5, 1861, they found that at a height of 21,000 feet the temperature sank to -10·4°; at 26,000 feet to -15·2°; and at 39,000 feet the temperature was down to -16·0° C. At this height the rarefaction of the air was so great and the cold so intense that Mr. Glaisher fainted, and Mr. Coxwell's hands being rendered numb and useless by the cold, he was only able to bring about their descent in time by pulling the string of the safety valve with his teeth. Yet when they [Pg 76]attained this height they were far above all cloud or mist, and the Sun's rays fell full upon them. The Sun's rays had all the force that they had at the surface of the Earth, but in the rare atmosphere of seven miles above the Earth, the radiation from every particle not in direct sunlight was so great that while the right hand, exposed to the Sun, might burn, the left hand, protected from his direct rays, might freeze.

But gravity at the surface of Mars is much feebler than at the surface of the Earth, and in order to reach the level of half-pressure a Martian mountaineer would have to climb, not three and a third miles, but eight and three-quarter miles; that is to say, the distance to be ascended is in the inverse proportion of the force of gravity at the surface of the planet. The atmosphere of Mars, therefore, is much deeper than that of the Earth, and one great cause of precipitation here is much weakened there. A current of air heavily laden with moisture, if it encounters a range of mountains, is forced upwards, and consequently expands, owing to the diminished pressure. The expansion brings about a cooling, and from both causes the atmosphere is unable to retain as much water-vapour as it carried before. On Mars, the same relative expansion and cooling would only follow if the ascent were nearly three times as great, and the feeble force of gravity has its effect in another way; for just as a weight on Mars will only fall six feet in the first second as against sixteen on the Earth, so a dense and heavy column of air will fall with proportionate slowness and a light column ascend in the same languid manner. An ascending current on Mars would therefore take $1/0{\cdot}38 \times 1/0{\cdot}38 = 1/0{\cdot}145$, or seven times as long to attain the same relative expansion as on the Earth.

The winds of Mars are therefore sluggish, and precipitation is slight. So far at least it resembles

"The island valley of Avilion;
Where falls not hail, or rain, or any snow,
Nor ever wind blows loudly;"

and R. A. Proctor, acute and accurate writer on planetary physics as he was, fell into a mistake when he referred to Mars as being "hurricane-swept." There are no hurricanes on Mars; its fiercest winds can never exceed in violence what a sailor would call a "capful."

This holds good for Mars, but it also holds good for every planet where the force of gravity at the surface is relatively feeble. The greater the force of gravity the more active the atmospheric circulation, and more violent its disturbances; the feebler the action of gravity the more languid the circulation, and the slighter the disturbances.

The atmosphere of Mars is relatively deeper than that of the Earth, so that we, in observing the details of its surface, are looking down through an immense thickness of an obscuring medium. And yet the details of the surface are seen with remarkable distinctness; not as clearly indeed as we can see those of the Moon, but nearly so. For instance, the "canals" appear to have a breadth of from 15 to 20 miles, corresponding to 1⁄16th, and 1⁄12th, of a second of arc, at an average opposition. The oases, as a rule, are about 120 miles in diameter, that is to say about half a second of arc. These are extraordinarily fine details to be perceived and held, even if Mars had no atmosphere at all; it would certainly be impossible to detect them unless the atmosphere were exceedingly thin and transparent. For we must remember that, though our own atmosphere is a hindrance to our observing, yet the atmosphere of the planet into which we are looking is a greater hindrance still. Like the lace curtains of the window of a house, it is a much greater obstacle to looking inward than to looking outward, and as the perfect distinctness with which we see the Moon is a proof that it is practically without an atmosphere, so the great detail visible on Mars bears unmistakable testimony to the slightness of the atmospheric veil around that planet.

And when we turn again to the statistics of Mars, we see that this must inevitably be the case. Of two planets, one heavier than the other, it is not possible to suppose that the lighter should secure the greater proportional amount of atmosphere. With planets, as with persons, it is the most powerful that gets the lion's share: "to him that hath it is given, and from him that hath not is taken away even that which he seemeth to have." But if we assume that Mars has acquired an atmosphere proportional to its mass, then we see from the Table that this must be a little less than 1⁄9th of that of the Earth; exactly 0·107. It is distributed over a smaller surface, 0·285. Consequently the amount of air above each square inch of Martian surface is 0·107 ÷ 0·285 = 0·38. But since the force of gravity at the surface of Mars is less than on the Earth, this column of air will only weigh 0·38 × 0·38 = 0·145; or one-seventh of the column of air resting on a square inch of the Earth's surface. The pressure at the surface of Mars will therefore be 2·1 lb.; and the aneroid barometer would read 4·3 inches. (In order to express the diminished pressure of the Martian atmosphere, it is necessary to refer it to the aneroid barometer. The mercury in a mercurial barometer, or the water in a water barometer would lose in weight in consequence of the diminished force of gravity in the same proportion as the air would, and the mercurial barometer would read 11·4 inches.)

But a pressure of 2·1 lb. on the square inch is far less than that experienced by Coxwell and Glaisher in their great ascent; it is about one-half the pressure that is experienced on the top of the very highest terrestrial mountains. But the habitable[Pg 80] regions of the Earth do not extend even so far upward as to the level of a pressure of 7·3 lb. on the square inch; that is, of half the terrestrial surface pressure. Plant life dies out before we reach that point, and though birds or men may occasionally attain greater heights, they cannot domicile there, and are, indeed, only able thus to ascend in virtue of nourishment which they have procured in more favoured regions. If we could suppose the conditions of the whole Earth changed to correspond with those prevailing at the summit of Mt. Everest, or even at the summit of Mont Blanc, it is clear that the life now present on this planet would be extinguished, and that speedily. Much more would this be the case if the atmosphere were diminished to one half the pressure on the summit of the highest earthly mountain.

The tenuity of the atmosphere on Mars has another consequence. Here water freezes at 0° C. and boils at 100° C.; so that for one hundred degrees it remains in a liquid condition. On Mars, under the assumed conditions, water would boil at 53° C., and the range of temperature within which it would be liquid would be much curtailed. But it is only water in the liquid state that is useful for sustaining life.

The above estimate of the density of the atmosphere of Mars is an outside limit, for it assumes that Mars has retained an atmosphere to the full proportion of its mass. But as the molecules of a[Pg 81] gas are in continual motion, and in every direction, the lighter, most swiftly moving molecules must occasionally be moving directly outwards from the planet at the top of their speed, and in this case, if the speed of recession should exceed that which the gravity of the planet can control, the particle is lost to the planet for ever. A small planet therefore is subject to a continual drain upon its atmosphere, a drain of the lightest constituents. Hence it is, no doubt, that free hydrogen is not a constituent of the atmosphere of the Earth.

To what extent, then, has the atmosphere of Mars fallen below its full proportion? Mr. Lowell has adopted an ingenious method of obtaining some light on this question, by comparing the relative albedoes of the Earth and Mars; that is to say the relative power of reflection possessed by the two planets. Of course the method is rough; we have first of all no satisfactory means of determining the albedo of the Earth itself, and Mr. Lowell puts it higher than most astronomers would do; then there is the difficulty of determining what portion of the total albedo is to be referred to the atmosphere and what to the actual soil or surface of the planet. But, on the whole, Mr. Lowell concludes that the amount of atmosphere above the unit of surface of Mars is 0·222 of that above the unit of surface of the Earth. This would bring down the pressure on each square inch of Mars to 1·2 lb., and the aneroid barometer would[Pg 82] read 2·5 inches; and water would boil at 44° C. The range of temperature from day to night, from summer to winter, at any place on the planet would be increased, while the range within which water could retain its liquid form would be diminished.

These statistics may seem rather dull and tiresome, but if we are to deal with the problem before us at all, it is important to understand that one factor in the condition of a planet cannot be altered and all the other factors retained unchanged. It will be seen that in computing the density of the atmosphere of Mars, we had to take into consideration not only the diameter of the planet, but the surface, which varies as the square of the diameter; the volume, which varies as the cube; the mass, which varies in a higher power still; and various combinations of these numbers. Novelists who write tales of journeys to other worlds or of the inhabitants of other worlds visiting this one, usually assume that the atmosphere is of the same density on all planets, and the action of gravity unchanged. In their view it is only that men would have a little less ground to walk upon on Mars, and a good deal more on Jupiter. Dean Swift, in *Gulliver's Travels*, made the Lilliputians take a truer view of the effect of the alteration of one dimension, for, finding that Gulliver was twelve times as tall as the average Lilliputian, they did not appoint him the rations of twelve Lilliputians, which would have been rather poor feeding for that veracious mariner, but allotted him the cube of twelve, viz. seventeen hundred and twenty-eight rations. Mr. J. Holt Schooling, in one of his ingenious and interesting statistical papers, tried to bring home the vast extent of the British Empire by supposing that it seceded, and taking the portion of Earth that has fallen to it, set up a world of its own—the planet "Victoria." He allots to the British Empire 21 per cent of the land surface of the world. If the Earth were divided so as to form two globes with surfaces in proportion of 21 to 79, the smaller globe, which would correspond to Mr. Schooling's new planet "Victoria," would be less than half the present Earth in diameter; it would be considerably smaller than Mars. But "the rest of the world" would be 0·96 of the present Earth in diameter, or very nearly the size of Venus, and it would contain just eight-ninths of the substance of the Earth, leaving only one-ninth for "Victoria." The statistics given above will suggest to the reader that, could such a secession be carried out, the inhabitants of the British Empire would not be happier for the change during the very short continued existence that remained to them. The "rest of the world" could spare our fraction of the planet much better than we could spare theirs.

This is a principle which applies to worlds [Pg 84]anywhere; not merely within the limits of the solar system but wherever they exist. Everywhere the surface must vary with the square of the diameter; the volume with the cube; everywhere the smaller planet must have the rarer atmosphere, and with a rare atmosphere the extreme range of temperature must be great, while the range of temperature within which water will flow will be restricted. Our Earth stands as the model of a world of the right size for the maintenance of life; much smaller than our Earth would be too small; much larger, as we shall see later, would be too large.

So far we have dealt with Mars as if it received the same amount of light and heat from the Sun that the Earth does. But, as the Table shows, from its greater distance from the Sun, Mars receives per unit of surface only about three-sevenths of the light and heat of that received by the Earth.

The inclination of the axis of Mars is almost the same as that of the Earth, so that the general character of the seasons is not very different on the two planets, and the torrid, temperate, and frigid zones have almost the same proportions. The length of the day is also nearly the same for both, the Martian day being slightly longer; but the most serious factor is the greater distance of Mars, and the consequent diminution in the light and heat received from the Sun. The light and heat received by the Earth are not so excessive that we could be content to see them diminished, even by 5 per cent, but for Mars they are diminished by 57 per cent. How can we judge the effect of so important a difference?

The mean temperature of our Earth is supposed to be about 60°F., or 16°C. Three-sevenths of this would give us 7°C. as the mean temperature of Mars, which would signify a planet not impossible for life. But the zero of the Centigrade scale is not the absolute zero; it only marks the freezing-point of water. The absolute zero is computed to be -273° on the Centigrade scale; the temperature of the Earth on the absolute scale therefore should be taken as 289°, and three-sevenths of this would give 124° of absolute temperature. But this is 149° below freezing-point, and no life could exist on a planet under such conditions.

But the mean temperature of Mars cannot be computed quite so easily. The hotter a body is the more rapidly it radiates heat; the cooler it is the slower its radiation. According to Stefan's Law, the radiation varies for a perfect radiator with the 4th power of the absolute temperature; so that if Mars were at 124° abs., while the Earth were at 289° abs., the Earth would be radiating its heat nearly 30 times faster than Mars. The heat income of Mars would therefore be in a much higher proportion than its expenditure; and necessarily its heat capital would increase until income and expenditure[Pg 86] balanced. Prof. Poynting has made the temperature of the planets under the 4th power law of radiation the subject of an interesting enquiry, and the figures which he has obtained for Mars and other planets are included in the Table.

The equatorial and average temperatures are given under the assumption that Mars possesses an atmosphere as efficient as our own in equalizing the temperature of the whole planet. If, on the other hand, its atmosphere has no such regulating power, then under the zenith Sun the upper limit of the temperature of a portion of its surface reflecting one-eighth would be, as shown in the Table, 64°C. This would imply that the temperature on the dark side of the planet was very nearly at the absolute zero. "If we regard Mars as resembling our Moon, and take the Moon's effective average temperature as 297° abs., the corresponding temperature for Mars is 240° abs., and the highest temperature is four-fifths of 337° = 270° abs. But the surface of Mars has probably a higher coefficient of absorption than the surface of the Moon—it certainly has for light—so that we may put his effective average temperature, on this supposition, some few degrees above 240° abs., and his equatorial temperature some degrees higher still. It appears as exceedingly probable, then, that whether we regard Mars as like the Earth or, going to the other extreme, as like the Moon, the[Pg 87] temperature of his surface is everywhere below the freezing-point of water." As the atmospheric circulation on Mars must be languid, and the atmosphere itself is very rare, the general condition of the planet will approximate rather to the lunar type than to the terrestrial, and the extremes, both of heat and cold, will approach those which would prevail on a planet without a regulating atmosphere.

There is another way of considering the effect on the climate of Mars and its great distance from the Sun, which, though only rough and crude, may be helpful to some readers. If we take the Earth at noonday at the time of the equinox, then a square yard at the equator has the Sun in its zenith, and is fully presented to its light and heat. But, as we move away from the equator, we find that each higher latitude is less fully presented to the Sun, until, when we reach latitude 64½°—in other words just outside the Arctic Circle—7 square yards are presented to the Sun so as to receive only as much of the solar radiation as 3 square yards receive at the equator. We may take, then, latitude 64½° as representing Mars, while the equator represents the Earth. Or, we may take it that we should compare the climate of Archangel with the climate of Singapore.

[Pg 88]Now the mean temperature of latitude 64½°, say the latitude of Archangel, is just about freezing-point (0°C.), while that of the equator is about 28°C. We should therefore expect from this a difference between the mean temperatures of the Earth and Mars of 28°; that is to say, as the Earth stands at 16°C, Mars would be at -12°C. But, on the Earth, the evaporation and precipitation is great, and the atmospheric circulation vigorous. Evaporation is always going on in equatorial regions, and the moisture-laden winds are continually moving polewards, carrying with them vast stores of heat to be liberated as the rain falls. The oceanic currents have the same effect, and how great the modification which they introduce may be seen by comparing the climates of Labrador and Scotland. There appear to be no great oceans on Mars. The difference of 28° which we find on the Earth between the equator and the edge of the Arctic Circle is a difference which remains after the convection currents of air and sea have done much to reduce the temperature of the equator and to raise that of high latitudes. If we suppose that their effect has been to reduce this difference to one half of what it would have been were each latitude isolated from the rest, we shall not be far wrong, and we should get a range of 56° as the true equivalent difference between the mean temperatures of Singapore and Archangel; i.e. of the Earth and Mars; and Mars would stand at -40°C. The closeness with which this figure agrees with that reached by Prof. Poynting suggests that it is a fair approximation to the correct figure.

The size of Mars taught us that we have in it a planet with an atmosphere of but one half the density of that prevailing on the top of our highest mountain; the distance of Mars from the Sun showed us that it must have a mean temperature close to that of freezing mercury. What chance would there be for life on a world the average condition of which would correspond to that of a terrestrial mountain top, ten miles high and in the heart of the polar regions? But Mars in the telescope does not look like a cold planet. As we look at it, and note its bright colour, the small extent of the white caps presumed to be snow, and the high latitudes in which the dark markings—presumed to be water or vegetation—are seen, it seems difficult to suppose that the mean temperature of the planet is lower than that of the Earth. Thus on the wonderful photographs taken by Prof. Barnard in 1909, the Nilosyrtis with the Protonilus is seen as a dark canal. Now the Protonilus is in North Lat. 42°, and on the date of observation—September 28, 1909—the winter solstice of the northern hemisphere of Mars was just past. There would be nothing unusual for the ground to be covered with snow and the water to be frozen in a corresponding latitude if in a continental situation on the Earth. Then, again, in the summer, the white polar caps of Mars diminish to a far greater extent than the snow and ice caps of the Earth; indeed, one of the Martian caps has been known to disappear completely.

Yet, as the accompanying diagram will show, something of this kind is precisely what we ought to expect to see. The diagram has been constructed in the following manner: A curve of mean temperatures has been laid down for every 10° of latitude on the Earth, derived as far as possible from accepted isothermals in continental countries in the northern hemisphere. From this curve ordinates have been drawn at each 10°, upward to show average deviation from the mean temperature for the hottest part of the day in summer, downward for the deviation for the coldest part of the night in winter. Obviously, on the average, the range from maximum to minimum will increase from the equator to the poles. The mean temperature of the Earth has been taken as 16°C, and as representing that prevailing in about 42° lat. The diagram shows that the maximum temperature of no place upon the Earth's surface approaches the boiling-point of water, and that it is only within the polar circle that the mean temperature is below freezing-point. Water, therefore, on the Earth must be normally in the liquid state.

In constructing a similar diagram for Mars, three modifications have to be made. First of all, the mean temperature of the planet must be considerably lower than that of the Earth. Next, since the atmospheric circulation is languid and there are no great oceans, the temperatures of different latitudes cannot be equalized to the same extent as on the Earth. It follows, therefore, that the range in mean temperature from equator to pole must be considerably greater on Mars than on the Earth. Thirdly, the range in temperature in any latitude, from the hottest part of the day in summer to the coldest part of the night in winter, must be much greater than with us; partly on account of the very slight density of the atmosphere, and partly on account of the length of the Martian year.

We cannot know the exact figures to adopt, but the general type of the thermograph for Mars as compared with that of the Earth will remain. The mean temperature of Mars will be lower, the range of temperature from equator to pole will be greater, and the extremes of temperature in any given latitude more pronounced than upon the Earth. And the general lesson of the diagram may be summed up in a sentence. The maximum temperature on the planet is well above freezing-point, and the part of the planet at maximum temperature is precisely the part that we see the best. But while this is so, it is clear that water on Mars must normally be in the state of ice; Mars is essentially a frozen planet; and the extremes of cold experienced there, not only every year but[Pg 93] every night, far transcend the bitterest extremes of our own polar regions.

The above considerations do not appear to render it likely that there is any vegetation on Mars. A planet ice-bound every night and with its mean temperature considerably below freezing-point does not seem promising for vegetation. If vegetation exists, it must be of a kind that can pass through all the stages of its life-history during the few bright hours of the Martian day. Every night will be for it a winter, a winter of undescribable frost, which it could only endure in the form of spores. So if there be vegetation it must be confined to some hardy forms of a low type. At a distance of forty millions of miles it is not easy to discriminate between the darkness of sheets of water and the darkness of stretches of vegetation. Some of the so-called "seas" may possibly be really of the latter class, but that there must be expanses of water on the planet is clear, for if there were no water surfaces there would be no evaporation; and if there were no evaporation from whence could come the supply of moisture that builds up the winter pole cap?

The great American astronomer, Prof. Newcomb, gave in *Harper's Weekly* for July 25, 1908, an admirable summary of the verdict of science as to the character of the meteorology of Mars. "The most careful calculation shows that if there are any considerable bodies of water on our neighbouring planet they exist in the form of ice, and can never be liquid to a depth of more than one or two inches, and that only within the torrid zone and during a few hours each day.... There is no evidence that snow like ours ever forms around the poles of Mars. It does not seem possible that any considerable fall of such snow could ever take place, nor is there any necessity of supposing actual snow or ice to account for the white caps. At a temperature vastly below any ever felt in Siberia, the smallest particles of moisture will be condensed into what we call hoar frost, and will glisten with as much whiteness as actual snow.... Thus we have a kind of Martian meteorological changes, very slight indeed and seemingly very different from those of our earth, but yet following similar lines on their small scale. For snowfall substitute frostfall; instead of feet or inches say fractions of a millimetre, and instead of storms or wind substitute little motions of an air thinner than that on the top of the Himalayas, and we shall have a general description of Martian meteorology."

What we know of Mars, then, shows us a planet, icebound every night, but with a day temperature somewhat above freezing-point. As we see it, we look upon its warmest regions, and the rapidity with which it is cleared of ice, snow, and cloud shows the atmosphere to be rare and the moisture little in amount and readily evaporated. The seas are probably shallow depressions, filled with[Pg 95] ice to the bottom, but melted as to their surfaces by day. From the variety of tints noted in the seas, and the recurrent changes in their outlines, they are composed of congeries of shallow pools, fed by small sluggish streams; great ocean basins into which great rivers discharge themselves are quite unknown.

CHAPTER VIII

THE ILLUSIONS OF MARS

The two preceding chapters have led to two opposing, two incompatible conclusions. In Chapter VI, a summary was given of Prof. Lowell's claim to have had ocular demonstration of the handiwork of intelligent organisms on Mars. In Chapter VII, it was shown that the indispensable condition for living organisms, water in the liquid state, is only occasionally present there, the general temperature being much below freezing-point, so that living organisms of high development and more than ephemeral existence are impossible.

Prof. Lowell argues that the appearance of the network of lines and spots formed by the canals and oases, and its regular behaviour, "preclude its causation on such a scale by any natural process," his assumption being that he has obtained finality in his seeing of the planet, and that no improvement in telescopes, no increase in experience, no better eyesight will ever break up the perfect regularity of form and position, which he gives to the canals, into finer and more complex detail.

But the history of our knowledge of the planet's surface teaches us a different lesson. Two small[Pg 97] objects appear repeatedly on the drawings made by Beer and Mädler in 1830; these are two similar dark spots, the one isolated, the other at the end of a gently curved line. Both spots resemble in form and character the oases of Prof. Lowell, and the curved line, at the termination of which one of the spots appears, represents closely the appearance presented by several of the canals. In the year 1830 no better drawings of Mars had appeared; and in representing these two spots as truly circular and the curved line as narrow, sharp, and uniform, Beer and Mädler undoubtedly portrayed the planet as actually they saw it. The one marking was named by Schiaparelli the Lacus Solis, the other, the Sinus Sabæus, and they are two of the best known and most easily recognized of the planet's features; so that it is easy to trace the growth of our knowledge of both of them from 1830 up to the present time. They were drawn by Dawes in 1864, by Schiaparelli in

1877 and the succeeding years, by Lowell in 1894 and since, and by Antoniadi in 1909 and 1911. But whereas the drawings of Beer and Mädler, made by the aid of a telescope of 4 inches aperture, show the two spots as exactly alike, in those of Dawes, made with a telescope of 8 inches, the resemblance between the two has entirely vanished, and neither is shown as a plain circular dot. Since then, observers of greater experience and equipped with more powerful instruments have directed their[Pg 98] attention to these two objects, and a mass of complicated structure has been brought out in the regions which were so simple in the sight of Beer and Mädler, so that not a trace of resemblance remains between the two objects that to them appeared indistinguishable.

Now the gradation in size, from the Lacus Solis down to the smallest oasis of Lowell, is a complete one. If a future development in the power of telescopes should equal the advance made from the 4-inch of Beer and Mädler, to the 33-inch which Antoniadi used in 1909, is it reasonable to suppose that Prof. Lowell's oases will refuse to yield to such improvement, and will all still show themselves as uniform spots, precisely circular in outline? It is clear that Beer and Mädler would have been mistaken if they had argued that the apparently perfect circularity of the two oases which they observed proved them to be artificial, because the increase in telescopic power has since shown us that neither is circular. The obvious reason why they appeared so round to Beer and Mädler was that they were too small to be defined in their instruments; their minor irregularities were therefore invisible, and their apparent circularity covered detail of an altogether different form.

Beer and Mädler only drew two such spots; Lowell shows about two hundred. Beer and Mädler's two spots seemed to them exactly alike; these two spots as we see them to-day have no [resemblance to each other. Prof. Lowell's two hundred oases, with few exceptions, seem all of the same character; is it possible to suppose, if telescopes develop in the future as they have done in the past, that the two hundred oases will preserve their uniformity of appearance any more than the Lacus Solis and the head of the Sinus Sabæus? If a novice begins to work upon Mars with a small telescope, he will draw the Lacus Solis and the Sinus Sabæus as two round, uniform spots, and as he gains experience, and his instrumental power is increased, he will begin to detect detail in them, and draw them as Dawes and Schiaparelli and others have shown them later. It is no question of planetary change; it is a question of experience and of "seeing."

There is a much simpler explanation of the regularity of the canals and oases than to suppose that an industrious population of geometers have dug them out or planted them; it is connected with the nature of vision.

A telegraph wire seen against a background of a bright cloud can be discerned at an amazing distance—in fact, at 200,000 times the breadth of the wire; a distance at which the wire subtends a breadth of a second of arc. For average normal sight the perception of the wire will be quite unmistakable, but at the same time it would be quite untrue to say that the perception of the wire was of the nature of defined vision, as would be seen at once if small objects of irregular shape were threaded on the wire; these would have to be many times the breadth of the wire in order to be detected. Again, if instead of a wire of very great length extending right across the field of view of both eyes, a short, black line be drawn on a white ground, it will be found that as the length of the line is diminished below a certain point so its breadth must be increased. If the observer is distant from the line 6000 times its length, then the breadth must be increased to be equal to the length, and the object, whatever its actual shape, can be just recognized as a small circular spot, which will subtend about 34 seconds of arc.

But though a black spot, 34 seconds in diameter, can be perceived on a white ground, we have not yet attained to defined vision. For if we place two black spots each 34 seconds of arc in diameter, near each other, they will not be seen as separate spots unless there is a clear space between them of six times that amount. Nearer than that they will give the impression that they form one circular spot, or an oval one, or even a uniform straight line, according to the amount of separation. If two equal round spots be placed so that the distance between their centres is equal to two diameters, then the diameter of each spot must be, at least, 70 seconds of arc for them to be distinctly defined; that is to say for the spots to be seen as two separate objects.

It will be seen that there is a wide range between objects that are large enough to be quite unmistakably perceived, and objects which are large enough to have their true outline really defined. It is a question of seconds of arc in the one case and of minutes of arc in the other. Within this range, between the limit at which objects can be just perceived and that where they can be just defined, objects must all appear as of one of two forms—the straight line and the circular dot.

This depends upon the structure of the eye and of the retina; the eye being essentially a lens with its defining power necessarily limited by its aperture, and the retina a sensitive screen built up of an immense number of separate elements each of which can only transmit a single sensation. Different eyes will have different limits, both for the smallest objects which can be discerned and for the smallest objects that can be defined, but for any sight the range between the two will be of the order just indicated.

Prof. Lowell has drawn attention to the "strangely economic character of both the canals and oases in the matter of form." It is true that straight lines and circles are economic forms, but they are economic not only in the construction of irrigation works but also in vision. "The circle is the figure which encloses the maximum area for the minimum average distance from its centre to any point situated within it;"therefore, if a small spot be perceived by the sight but be too small to have its actual outline defined, it will be recognized by the eye as being truly circular, on the principle of economy of effort. So, again, a straight line is the shortest that can be drawn between two points; and a straight line can be perceived as such when of an angular breadth quite 40 times less than that of the smallest spot. A straight line is that which gives the least total excitement in order to produce an appreciable impression, and therefore the smallest appreciable impression produces the effect of a straight line.

It is sufficient, then, for us to suppose that the surface of Mars is dotted over with minute irregular markings, with a tendency to aggregate in certain directions, such as would naturally arise in the process of the cooling of a planet when the outer crust was contracting above an unyielding nucleus. If these markings are fairly near each other it is not necessary, in order to produce the effect of "canals," that they should be individually large enough to be seen. They may be of any conceivable shape, provided that they are separately below the limit of defined vision, and are sufficiently sparsely scattered. In this case the eye inevitably sums up the details (which it recognizes but cannot resolve) into lines essentially "canal-like" in character. Wherever there is a small aggregation of these minute markings, an impression will be given of a circular spot, or, to use Prof. Lowell's nomenclature, an "oasis." If the aggregation be greater still and more extended, we shall have a shaded area—a "sea."

The above remarks apply to observation with the unaided eye, but the same principle applies yet more strongly to telescopic vision. No star is near enough or sufficiently large to give the least impression of a true disc; its diameter is indistinguishable; it is for us a mathematical point, "without parts or magnitude." But the image of a star formed by a telescope is not a point but a minute disc, surrounded by a series of diffraction rings. This disc is "spurious," for the greater the aperture of the telescope the smaller the apparent disc.

That which holds good for a bright point like a star holds good for every individual point of a planetary surface when viewed through the telescope; that is to say, each point is represented by a minute disc; all lines and outlines therefore are slightly blurred, so that minute irregularities are inevitably smoothed out.

When we come to photographs, the process is carried to a third stage. The image is formed by the telescope, subject to all the limitations of telescopic images, and is received on a plate essentially granular in structure, and is finally examined by the eye. The granular structure of the plate acts as the third factor in concealing irregularities and simplifying details; a third factor in producing the two simplest types of form—the straight line and the circular dot.

Prof. Lowell describes the canals as like lines drawn with pen, ink and ruler, but not a few of our best observers have advanced much beyond this stage. Even as far back as 1884, some of the canals were losing their strict rectilinear appearance to Schiaparelli, and the observers of the planet who have been best favoured by the power of the telescope at their disposal, by the atmospheric conditions under which they worked, and by their own skill and experience—such as Antoniadi, Barnard, Cerulli, Denning, Millochau, Molesworth, Phillips, Stanley Williams and others—have found them to show evident signs of resolution. Thus, in 1909, Antoniadi found that of 50 canals, 14 were resolved into disconnected knots of diffused shadings, 4 were seen as irregular lines, 10 as more or less dark bands; and he found that, in good seeing, there was no trace whatever of the geometrical network.

The progress of observation, therefore, has left Prof. Lowell behind, and has dispelled the fable which he has defended with so much ingenuity. But, indeed, there never was any more reason for taking seriously his theory as to the presence of artificial waterways on Mars than for believing in the actual existence of the weird creatures described by H. G. Wells in the *War of the Worlds*.

There are too many oversights in the canal theory.

Thus no source is indicated for the moisture supposed to be locked up in the winter pole cap. Prof. Lowell holds that there are no large bodies of water on the planet; that the so-called seas are really cultivated land. In this case there could be little or no evaporation, and so no means by which the polar deposits could be recruited.

Yet it is certain that the supply of the winter pole cap must come from the evaporation of water in some region or other. And here is another oversight of the artificial canal theory. The canals are supposed to be necessary for the conveyance of water from the pole towards the equator; although, as this was "uphill," vast pumping stations at short intervals had to be predicated. But it is not supposed that the water needed to travel by way of the canals to the poles. If, however, the moisture is conveyed as vapour through the atmosphere to the pole as winter approaches, it cannot be impossible that it should be conveyed in the same manner from the pole as summer draws on, and in that case the artificial canals would not be needed. If the canals are necessary for conveying the water in one direction, they would be necessary for the opposite direction. But there would be something too farcical in the idea of the careful Martians dispatching their water first to the pole to be frozen there, and then, after it had been duly frozen and melted again, fetching it back along thousands of miles and through numerous pumping stations for use in irrigating their fields.

Of all the many hundreds of canals only a few actually touch the polar caps. But on the theory that the entire canal system is fed by the polar cap in summer, the carrying capacity of the polar canals should be equal to, if not greater than, that of the entire system outside the polar circle. A glance at the charts of the planet shows that the polar canals could not supply a twentieth part of the water needed for those in the equatorial regions. Another oversight is that of the significance of the alleged uniformity and breadth of the canals. Prof. Lowell repeatedly insists that the canals are of even breadth from end to end, and spring into existence at once throughout their whole length. This statement is in itself a proof that the canals cannot be what he supposes them to be. An irrigation system could not have these characteristics; the region fertilized would take time to develop; we should see the canal extending itself gradually across the continent, and its breadth would not be uniform from end to end, but the region fertilized would grow narrower with increase of distance from the fountain head of the canal.

Under what conditions can we see straight lines, perfectly uniform from end to end, spring into existence, in their entirety, without going through any stages of growth? When the lines are not actual images, but are suggested by markings perceived, but not perfectly defined. In 1902 and 1903, in conjunction with Mr. Evans, the headmaster of Greenwich Hospital School, I tried a number of experiments on this point, with the aid of about two hundred of the boys of the school. They had several qualifications in respect of these experiments; they were keen-sighted, well drilled; accustomed to do what they were told without asking questions; and they knew nothing whatsoever of astronomy, certainly nothing about Mars.

A diagram was hung up, based upon some drawing or other of the planet made by Schiaparelli, Lowell or other Martian observer, but the canals were not inserted; only a few dots or irregular markings were put in here and there. And the boys were arranged at different distances from the diagram and told to draw exactly what they saw.

Those nearest the diagram were able to detect the little irregular markings and represented them under their true forms. Those at the back of the room could not see anything of them, and only represented the broadest features of the diagram, the continents and seas. Those in the middle of the room were too far off to define the minute markings, but were near enough for those markings to produce some impression upon them; and that impression always was of a network of straight lines, sometimes with dots at the points of meeting. Advancing from a distance toward the diagram the process of development became quite clear. At the back of the room no straight lines were seen; as the observer came slowly forward, first one straight line would appear completely, then another, and so on, until all the chief canals drawn by Schiaparelli and Lowell in the region represented had come into evidence in their proper places. Advancing still further, the canals disappeared, and the little irregular markings which had given rise to them were perceived in their true forms.

These experiments at the Greenwich Hospital School were merely the repetition of similar ones that I had myself made privately twelve years earlier, leading me to the conclusion, published in 1894, that the canals of Mars were simply the summation of a complexity of detail too minute to be separately discerned.

A little later, in his work *"Marte nel 1896-7,"* Dr. Cerulli independently arrived at the same conclusion, and wrote: "These lines are formed by the eye ... which utilizes ... the dark elements which it finds along certain directions"; and "a large number of these elements forms a broad band"; and "a smaller number of them gives rise to a narrow line." Also, "the marvellous appearance of the lines in question has its origin, not in the reality of the thing, but in the inability of the present telescope to show faithfully such a reality." In 1907, Prof. Newcomb made some experiments in the same direction and reached the same general conclusion. More recently still, Prof. W. H. Pickering has worked on the same lines and with the same result. The venerable George Pollock, formerly the Senior Master of the Supreme Court and King's Remembrancer, sent to me, in his 91st year, the following note as affording an apt illustration of the true nature of the canaliform markings on Mars:

"On Saturday last, journeying in a motor-car, I came into a broad road bounded by a dark wood. Looking up I was amazed to see distinct, well-defined, vertical, parallel white lines, the wood forming the dark background. On getting nearer, these lines resolved themselves into spots, and they proved to be the white insulators supporting the telegraph wires."

Prof. Lowell has objected that all experiments and illustrations of this kind are irrelevant; only observations upon the planet itself ought to be taken into account.

But such observations have been made upon the planet itself with just the same result. Observers have seen streaks upon Mars—knotted, broken, irregular, full of detail—and when the planet has receded to a greater distance, the very same marking has shown itself as a narrow straight line, uniform from end to end, as if drawn with pen, ink and ruler. The greater distance has caused the irregularities, seen when nearer at hand, to disappear. In this, and not in any gigantic engineering works, is the explanation of the artificiality of the markings on Mars as Prof. Lowell sees them. That artificiality has already disappeared under better seeing with more powerful telescopes.

This chapter is entitled "The Illusions of Mars." Yet the illusions of Mars are not the straight lines and round dots of the canal system, but the forced and curious interpretation which has been put upon them. If the planet be within a certain range of distance and under examination with a certain telescopic power, the straight lines and round dots are inevitable. Their artificiality is not a function of the actual Martian details themselves, but of the mode in which, under given conditions, we are obliged to see them.

CHAPTER IX

VENUS, MERCURY AND THE ASTEROIDS

Of all the planets, Venus appears, to the unassisted eye, by far the loveliest. When seen in the early morning before sunrise—its "western elongation"—or after sundown in the evening—its "eastern elongation"—and still more as it attains its greatest brilliancy, it has attracted attention everywhere and in all ages. It then shines with brilliance ten times as great as Jupiter in opposition, and the brightest members of the heavenly host look pale and dim beside it. It is emphatically the morning or the evening star, Lucifer, or Vesper, herald or follower of the Sun; it can even assert itself in the presence of the Lord of Day, for it has often been seen at noonday by watchers who knew where to look; sometimes by the general crowd.

But in the telescope Venus appears less satisfying. It is a pretty spectacle indeed to watch the phases of the gleaming little globe of silver, for, like the Moon under varying illumination from the Sun, it undergoes change of apparent shape. But the surface of the planet yields little detail, and that little is illusive and ill-defined. The clear-cut outlines and black shadows of the Moon have no place here, nor do the ruddy plains and blue-grey "seas" of Mars find any analogues. All that can be observed beyond the changes of phase are a few faint, ill-defined patches, where the molten silver of the general surface is slightly dimmed and tarnished, and perhaps one or two spots, not less evasive and difficult to fix, that exceed the rest of the surface in brightness.

This very difficulty in making out the markings on Venus is hopeful for our search; it points to a veiling over the planet, a veiling by an atmosphere. And the statistics of the Table show that Venus closely resembles our Earth in size and mass, and therefore probably in atmospheric equipment. If we assume that the atmosphere of any planet is in direct proportion to its mass—and as Venus is so nearly the twin of the Earth there is no reason to expect any great difference between the two in this respect—the atmosphere of Venus would have a pressure of about 11·2 lb. on the square inch, and the level of half pressure would be nearly four miles above the surface. In other words the atmosphere would be both thinner and deeper than that of the Earth, but the difference would not be important in amount.

But Venus is nearer to the Sun than the Earth, and receives nearly double the light and heat. Its theoretical equatorial temperature is 368°abs., or 95°C, and its corresponding mean temperature is 69° C. But water under a pressure of 11·2 lb. will boil at 93° C, so that at the equator of Venus the upper limit for water as a liquid is just passed, but, for the planet in general, a fairly safe margin is maintained. Here then is sufficient explanation why the topography of Venus is concealed. The atmosphere will always be abundantly charged with water-vapour, and an almost unbroken screen of clouds be spread throughout its upper regions. Such a screen will greatly protect the planet from the full scorching of the Sun, and tend to equalize the temperature of day and night, of summer and winter, of equator and poles. The temperature range will be slight, and there will be no wide expanses of polar ice. Water that flows will be abundant everywhere.

So far all the facts connected with Venus are favourable for life, even though the picture called up to the mind may not seem inviting to us. For views of the heavens must be rare; the Sun must seldom pierce through the cloud veil; there is no moon and the stars must be almost always hidden. The Earth with its Moon might form a beautiful ornament at times in the midnight sky if the cloud-shell should occasionally open, but on the whole, the planet is shut up to itself in a perpetual vapour-bath, and its condition will approach that of some of the most humid countries in the terrestrial tropics during the height of their rainy seasons.

But it would seem that life both of plants and animals, under such conditions, might flourish and be abundant. The mean temperature would not, in general, be high enough to drive off the water as steam, nor low enough to congeal it into ice; it would remain water—water that flows.

But there is still a possible hindrance to life on Venus, a hindrance that actually exists in the case of Mercury.

Mercury, the "Twinkler," is not an easy object in our Northern latitudes, but, in countries near the tropics, is often quite conspicuous, a little scintillating gem of light in the bright sky, before sunrise or after sunset. In the telescope it is not so attractive as Venus, partly because it is smaller, partly because, though it receives more than three times as much light from the Sun, it is duller in hue. Yet it is not quite so secretive as its neighbour, and a certain number of markings have been detected upon its disc, markings which, like those of the Moon, appear to be permanent.

A glance at the Table will show that this was to be expected. In size, Mercury comes between the Moon and Mars, and the atmospheric veil ought therefore to be, as it evidently is, very slight and transparent; offering little or no hindrance to an observer scanning it from another world. The other necessary consequences of small size and mass will follow; the feeble force of gravitation, the languid atmospheric circulation, the extreme range of temperatures, the low temperature at which water will boil.

But the heat to which Mercury is exposed far transcends our terrestrial experience. In the mean it receives nearly seven times as much heat from the Sun as the Earth does, but this supply is not maintained uniformly, for Mercury moves round the Sun in a very eccentric orbit, so that when in aphelion it receives, surface for surface, only about four times as much heat as the Earth, but some six weeks later when in perihelion it receives more than eleven times. The great range of temperature due to the thinness of the atmosphere must therefore be further increased by the varying distance of the planet from the Sun.

A reference to Prof. Poynting's figures shows that the mean temperature of Mercury must approximate to 194° C., while water will boil at 40° C. or even lower. Here, then, is a condition the exact reverse of Mars. Water as a liquid will be rare on Mercury, not because it is congealed, but because it is evaporated; on the dark side of the planet it may, indeed, pass into ice, but on the side exposed to the Sun it must exist normally as a constituent of the atmosphere. Water in a liquid state, water that flows, must be almost unknown.

But we have good reason to believe that that which is the dark side of Mercury at one time is always dark; that which is exposed to the Sun is always exposed to it.

Since Mercury wears no concealing veil of atmosphere, and displays markings that can be identified and followed, a surprising circumstance has come to light. In 1889, Schiaparelli discovered that Mercury, instead of rotating on its axis in about 24 hours like the Earth and Mars, rotates in 88 days; that is to say, it always turns the same face towards the Sun, just as the Moon turns the same face towards the Earth. This fact, confirmed theoretically by Prof. G. H. Darwin in his development of the theory of tidal friction, puts the condition of Mercury in quite a new light. No alternation of day or night refreshes and restores the little world; one hemisphere is for ever exposed to the blasting heat of the Sun, seven times hotter for it than for the Earth; the other hemisphere is for ever exposed to the darkness and cold of outer space, a range from something like 390° C. above freezing-point, to 270° C. below. It is true that between the two hemispheres there is a "debatable land," for, owing to the ellipticity of the orbit, the face turned to the Sun is not exactly the same at all times, and a region about 47° in width on each side of the planet, that is to say, rather more than a quarter of its entire surface, has one day and one night in each period of 88 days, but these more favoured sections can scarcely be considered more habitable than the rest.

The conditions of Mercury are so unfavourable for life that, even if this remarkable relation of rotation period to revolution did not hold good, it would still be impossible to regard it as a world for habitation. But its case shows that a further condition of habitability has to be satisfied by a planet. Size and distance from the Sun afford the first two conditions; a suitable rotation period is now seen to be a third.

And it is possible that in this very particular Venus fails to qualify. Schiaparelli, the first observer of his time, assisted by the clear Italian sky, believed that he had demonstrated that Venus, like Mercury, rotates once in her year; her day being thus equal in length to 225 of ours, and the face that she turns to the Sun being always the same.

And in her case this statement requires practically no qualification, for, her orbit being nearly circular, there is hardly any libration; a place that has the Sun in its zenith has it so for ever; one on the night side of Venus can never have a sunrise, or gladden in the daylight. The side exposed to the Sun will wither in a temperature of about 227° C., in which all moisture will be evaporated; the side remote from it will be bound in eternal ice. In neither hemisphere will water exist in the liquid state; in neither hemisphere will life be possible.

But as yet the evidence is not conclusive that Venus has this long rotation period. Several observers of high rank believe that our neighbour rotates in nearly the same time as the Earth, but its markings are so faint and elusive that the problem is a difficult one. The spectroscopic method of determining the speed of rotation has been equally indecisive. Until, therefore, the rotation period has been decided, the habitability of Venus must remain in question. If it always turns the same face to the Sun, there can be no more life upon it than upon Mercury; if on the contrary it rotates in much the same time as the Earth, then, so far as we know, it may well be a habitable world. Whether it is actually inhabited is a matter at present entirely beyond our knowledge.

A page or two back we touched lightly on the eccentricity of the orbit of Mercury—lightly, because it was not the chief factor in disabling the planet for habitation. But the condition introduced by this eccentricity is one which of itself would be sufficient to put it out of court. In the six weeks in which Mercury moves from aphelion to perihelion, it approaches the Sun by fourteen millions of miles, and the heat received by it is increased 2½ times. Then, in the next six weeks, it recedes as far, and there is a like diminution. In other words, six weeks makes a greater proportional change in this one planet's condition than we should experience if our Earth were transported from its own orbit to that of Mars.

But there are other members of the solar system whose orbits are so elongated that that of Mercury seems in comparison almost circular. These are the comets, some of which all but graze the surface of the Sun at perihelion, and then recede from him for periods that it takes even thousands of years to complete. But without dwelling on such extreme cases, two of the best known of the periodic comets may be taken as examples of the rest. Encke's is the comet of shortest period, returning in about 3·3 years. At perihelion it is 31 millions of miles from the Sun; one-third the distance of the Earth. It receives, therefore, at this part of its orbit, 9 times as much light and heat as the Earth. But at aphelion it retreats deep into the region of the asteroids, and is much more than four times the mean distance of the Earth. At this part of its orbit it receives but 1⁄17th as much heat as the Earth. By far the most famous of all the comets is that known by the name of Halley, and its mean period is 76 years. At perihelion it comes within the orbit of Venus; indeed, nearly halfway between that and the orbit of Mercury. At aphelion it recedes to thirty-five times the distance of the Earth, far beyond the orbit of Neptune. The range in its light and heat from the Sun is from 3 times that of the Earth to less than 1⁄1200th; or, in other words, the supply of heat at one time is nearly 4000 times that at another, and of the 76 years of its period, only 80 days are spent within the orbit of the Earth.

Comets cannot be homes of life; they are not sufficiently condensed; indeed, they are probably but loose congeries of small stones. But even if comets were of planetary size it is clear that life could not be supported on them; water could not remain in the liquid state on a world that rushed from one such extreme of temperature to another.

Between the orbits of Mars and Jupiter there are scattered an untold number of little planets commonly known as asteroids or minor planets. Minor planets indeed they are, for the one first discovered—Ceres—probably outweighs all the rest, known and unknown, put together, though something like 700 have already been detected, and the list grows at the rate of about one a week.

As the Table shows, Ceres is so small that the Earth exceeds it in volume 5000 times; even the Moon is 90 times as large. The mass of Ceres is not known; being so small, its density is probably less than that of the Moon, so that the Earth may easily outweigh it 10,000 times. The unfavourable conditions resulting from smallness of size that the Moon presents are therefore exaggerated exceedingly in the case of Ceres; its atmosphere must approach in tenuity what we should regard as a vacuum in a terrestrial laboratory, and water as a liquid be entirely unknown. Its distance from the Sun is another hostile factor; for in consequence it receives per unit of surface only 13 per cent of the light and heat that falls on the Earth; its maximum temperature under a zenith Sun will fall far below freezing-point, the minimum on the dark side will approach the absolute zero.

With Ceres the whole of the asteroidal family can be dismissed as possible abodes of life. No astronomer can regard them as such. Yet they have their lesson to teach. Life can exist on the Earth only on the upper face of its crust, and in a very thin film of air and water; but the enormous solid bulk within, inert though it be, that supports the stage on which the great drama of life is played, is as really essential as air and water themselves. If that bulk were much smaller and less massive life could find no place upon its surface.

CHAPTER X

THE MAJOR PLANETS

It is a striking change to pass from Ceres, the giant of the minor planets, to Jupiter, the giant of the major planets. Instead of a world that the Earth exceeds in volume 5000 times, we are confronted by one that exceeds the Earth 1400 times. Ceres, when viewed through a large telescope, is just able to present a perceptible disc; Jupiter offers the largest shown by any heavenly body after the Sun and Moon.

And that disc is one that never fails to charm the attentive student, for it abounds in colour, movement and change. The late Prof. James Keeler, an observer of the first rank, having the advantage of observing the planet from the summit of Mt. Hamilton and with the great 36-inch telescope of the Lick Observatory, thus describes the aspect of the planet in 1889.

"Seen with this instrument on a fine night, the disc of Jupiter was a most beautiful object, covered with a wealth of detail which could not possibly be accurately represented in a drawing.... Scarcely any portion of Jupiter, except the Red Spot and the extreme polar regions, was of a uniform tint, the surface being mottled with flocculent and more or less irregular cloud masses.... The equatorial zone, occupying the space between the red belts, was marked in the centre by a salmon-coloured stripe, which was occasionally interrupted by an extension of the white clouds on the sides of the zone. The edges were brilliant white, and were formed of rounded cloud-like masses, which at certain places extended into the red belts as long streamers.... Near their junction with the equatorial zone, the streamers were white and definite in outline, but they became redder in tint toward their outer extremities, and more diffuse, until they were lost in the general red colour of the background. When the seeing was good they were seen to be formed of irregular rounded or feathery clouds, fading toward the outer ends, until the structure could no longer be distinguished.... The portions of the equatorial zone surrounding the roots of well-marked streamers were somewhat brighter than at other places, and it is a curious circumstance that they were almost invariably suffused with a pale olive-green colour, which seemed to be associated with great disturbance, and which was rarely seen elsewhere.... The red belts presented on all occasions the appearance of a passive medium, in which the phenomena of the streamers and other forms ... were manifested. The phenomena would be exactly reproduced by streamers of cloudy white matter floating in a semi-transparent reddish fluid, sometimes submerged and sometimes rising to the surface.... The dark spots frequently seen on the red belts usually occupied spaces left by sharp turns in the streamers, and they were of the same colour as the belts, but deeper in tint, as if the fluid medium could be seen to a greater depth."

In other words, Jupiter is a striped or banded planet, the bands lying along the direction of turning. These bands are coloured in varying tints, and the planet rotates very rapidly, for the details in the bands pass quickly from one limb to the other. And not only is the speed of rotation of the whole very rapid—Jupiter turns about its axis in a little less than ten hours, so that a particle at its equator moves through 466 miles in each minute—but the various items that form the bands rotate in different times. They may also alter their form and their colour. Jupiter seems, then, to be a planet with a great and rapidly changing atmosphere that extends above a shoreless sea formed of some liquified substance or substances—the whole in a state of flux.

But if we turn back to the Table, we see that Jupiter at its mean distance from the Sun is 5·2 times that of the Earth; that is to say, it receives only 1/27th of the light and heat that we receive. But in Chapter VIII, we learnt from Mars that as this receives only 3/7ths of the Earth's light and heat, its mean temperature would sink to -30°C.; the Earth's being 16°C. Mars is therefore almost always a frozen planet; frozen except on its mere surface when this is exposed to the full rays of the Sun. No sea there would ever bemelted to a depth of more than a few inches, even at noonday in midsummer. And yet Mars has at least ten times the advantages of Jupiter. Jupiter, then, must be a frozen planet through and through; no liquid of any sort can exist on its surface; no vapour of any substance can exist in its atmosphere. It must be icebound even at its summer noonday.

Yet, from the description given by Prof. Keeler, it is manifestly not so; and another item in the Table emphasizes that it cannot be so. The density of the Sun is 1·4 that of water, Jupiter's is 1·33, showing that but a very small proportion (if any) of its bulk can be solid; the rest must be vaporous, or at least fluid. How then can we reconcile these inconsistencies?

It is in the dimensions of Jupiter that we find the answer. The mass of the planet is 317 times that of the Earth; it is indeed nearly three times as great as that of all the other planets put together. But the aggregation of so vast an amount of material is of itself a source of heat; the chief source at the present time of the enormous output of heat from the Sun is ascribed to its gradual contraction; the slow falling of its substance, if we may so express it, a little nearer to its centre. The great mass of Jupiter points to its inherent store of heat being much greater than that of any other planet. And of two bodies equally hot, the larger must cool more slowly than the smaller. If, therefore, all the members of the solar system had at one and the same moment possessed the same surface temperature, that equality would have ceased directly they began to radiate their heat into space; the temperature of the smaller bodies falling more rapidly than those of the larger. This is another example of the principle that has already been noted, that the properties of a small world are not those of a large world divided by a constant factor. It is not possible to conceive a model of the solar system in which all the significant factors should be true to the same scale. If the diameters and distances were all made on a one-tenth scale, the surfaces would be one-hundredth of reality, the volumes one-thousandth.

But a radiating body radiates from its surface, while the store of heat from which that radiation is kept up is supplied by its volume. It follows, therefore, that a large and heavy world must differ from a small light world, not merely in scale, but also in kind.

The surface of a world is all that we see of it; it is, therefore, very commonly all that we consider. But unseen, and hence often unconsidered, beneath the surface lies its substance or mass, and it is this that determines the state and condition of the surface; it is the underlying power. Two men may be contending in a financial struggle; to the eye they may look alike, equally prosperous; both may have the same amount of money actually in their pockets; but the one has nothing else, the other has a great banking account and vast investments, and is, in fact, a millionaire; and it is his unseen power and resources that will make themselves felt.

Jupiter therefore introduces us to a new factor in world-condition; not all its heat is derived from the Sun; much is inherent to it. And though it is not possible at present to say that the mass of Jupiter being so much its inherent heat must be this or that quantity as a function of that mass, yet in general, and neglecting other considerations, we can say that of two worlds the one with the greater mass will be that with the higher inherent temperature. This factor of inherent temperature was one that did not require to be noticed in dealing with the Moon, or Venus, or Mars, for these and all the planets yet noticed are less in size, surface, volume, and mass than the Earth, and hence possess less inherent heat. It is only now that the greater planets are being considered that the question of a source of heat, other than the Sun, can arise.

But the evidence of such heat on Jupiter is not to be disputed. The albedo or reflective index of Jupiter has been put by the late Prof. G. Bond, of Harvard College Observatory, as higher than unity; in other words, that it emits more light than it receives. This is now generally regarded as an excessive estimate, but the albedo of the disc as a whole cannot be put lower than 0·72, or about that of white paper. But many of the "belts" or dark regions are of a dull copper tint, and the polar caps are dusky, so that Bond's estimate must be realized for the most brilliant "zones," as the brighter regions are called; certainly for the whitest of the white spots.

No direct evidence of inherent luminosity has been obtained, for the satellites disappear entirely in eclipse. But though their shadows in transit appear very dark, it is clear that they are not absolutely black, since sometimes such a shadow is not distinguishable in darkness from the satellite that casts it; a delicate proof that the background on which it falls has some intrinsic luminosity.

Unless there is the counteracting effect of a high temperature, the atmosphere of Jupiter would have a pressure at the surface of 104 lb. to the square inch, and the level of half pressure be attained at a mile and a quarter; the reverse condition to that on Mars would obtain, and the atmosphere of Jupiter would be much denser and much shallower than that of the Earth. Denser it probably is; shallower it cannot be, for the great white spots, each often five or six thousand miles in diameter, that range themselves at times along the equatorial regions till they look like the portholes of a ship, evidently rise from depths great even as compared with their size. But it is only by intense heat that the effect of the great mass of Jupiter in constricting its atmosphere within shallow depths can be overcome.

Again, the extraordinary lightness of the planet, so little above the density of water, points in the same direction. So, not less unmistakably, do the magnitude and rapidity of the atmospheric movements. The clouds and storms of our own atmosphere are worked by solar heat; solar heat it is that draws up the vapours and provides the chief part of the energy manifested in the speed and strength of the air-current. But solar heat can only give 1/27th the amount of that energy at the distance of Jupiter, so that, if they were entirely dependent on solar radiation, the winds of Jupiter should be very feeble.

Further, the difference of presentment due to the difference of latitude is a fruitful cause of inequalities of temperature and pressure in the terrestrial atmosphere. But as a degree of latitude on Jupiter is eleven times as wide as on the Earth, such inequalities connected with a given difference in latitude are spread over eleven times the distance that they would be on the Earth, and are, therefore, so much the less pronounced. Yet, across a gulf of 400 millions of miles we can clearly discern the bright zones of Jupiter now narrowing down and constricting the red belts, now thrust apart by them, and can detect changes taking place in an hour of time over areas equal to that of a terrestrial hemisphere.

A notable peculiarity of Jupiter is found in the proper motions of its spots. Many of the white spots are exceedingly swift, giving a rotation period of 9h. 50m. while the equatorial belt in general gives a period 5m. longer; so that in 119 rotations (nearly 49 days) a white spot will have passed entirely round the belt, gaining upon it at a rate of nearly 240 miles an hour.

The most famous of all the markings in Jupiter is the Great Red Spot, which became conspicuous in 1878, since when the spot itself, or at least the nest in which it lay, has always been visible. It has been identified with a great red spot observed by Hooke and Cassini in 1664-6, that appeared and vanished again eight times between 1665 and 1708. It therefore has had a history practically as long as our telescopic knowledge of the planet, and may be looked upon as in some sort a permanent feature. Yet that it is not in the nature of a portion of a solid crust is clear. It occupies on Jupiter much the position and relative area of Australia on the Earth, but whereas Australia of necessity rotates in one piece with all the other continents, the Great Red Spot has a rotation period which is neither that of the equatorial belt, nor of the quickly moving white spots, and is not itself stable. An "Australia on the loose" is impossible, even unthinkable here, but the Great Red Spot, for all its long duration, is mobile and inconstant, and is therefore no portion of a solid permanent crust.

The giant planet Jupiter, therefore, offers us an example of what we may call a "semi-sun"; a world still bubbling with tremendous energies of its own, still pulsing with its own inherent heat, still without a solid crust; probably without a solid nucleus, liquid or vaporous throughout. Whatever the future may hold for such an orb, it is clearly no world for habitation at present. Full of colour, and movement, and change as it is, it lacks the Earth's "gloom of iron substance," which is necessary, no less than its veiling by the plant, as a stage for "the passion and perishing of mankind."

But if Jupiter be a semi-Sun, still a source of heat, perhaps even of light, can it yield the means of life to its satellites? For Jupiter is sun-like, not merely in its own condition, but also in that it is the centre and ruler of a system of its own. We know already of eight satellites revolving round it.

Of these eight, only four—the four discovered by Galileo, in the first days of his possession of a telescope—need be considered; the other four are of the same order of size as the asteroids, and are indeed much smaller than Ceres.

But the Galilean satellites are of a higher rank. Europa, the smallest, is in size a twin to the Moon; Callisto, the outermost, is almost exactly the size of Mercury; Io, the innermost, is midway between the two in its dimensions. But Ganymede, the largest, is almost comparable with Mars, its diameter being 0·45 that of the Earth instead of the 0·53 of Mars.

But the Moon, Mercury, and Mars have all been shown, on the ground of their small size, to be worlds unfit for habitation; the satellites of Jupiter are, therefore, all rejected on the same score. Nor can the greater nearness of their immediate primary compensate for their remoteness from the Sun. It is true that Jupiter presents to Ganymede a disc with more than 200 times the apparent area that the Sun presents to the Earth, but to make up for the falling-off of the solar radiation, each unit of this area should radiate about 1/250th as much heat as each unit of the Sun's surface. In other words, the absolute surface temperature of Jupiter should be ¼th that of the Sun, or about 1550° C., and this is higher than can be admitted. The Sun and Jupiter together cannot put Ganymede in as favourable a position as Mars, much less as favourable as the Earth.

The case of Jupiter carries with it those of Saturn, Uranus, and Neptune. All three, from their high albedoes and low densities, are still in a vaporous condition; still in some sort, semi-Suns; sources of a certain amount of heat, and not recipients merely. The days are yet far distant when a solid crust can form on any one of them, and the water condense from the steamy atmosphere to form oceans, seas, and rivers. Not till then, if at all, when water as a liquid, water that flows, is present, can life begin to appear and enter on its long course of change.

CHAPTER XI

WHEN THE MAJOR PLANETS COOL

The question has been asked: "It is evident that life cannot exist at the present time on the outer planets, since they are in a highly heated and quasi-solar condition; but when they cool down, as cool they must, and a solid crust is formed, may not a time come when they will be habitable? It seems impossible to think that worlds so beautiful to our eyes and so vast in scale are destined never to be peopled by intelligent beings."

It is clearly difficult to answer satisfactorily a question that requires so deep a plunge into the recesses of the unknown future; yet, so far as our knowledge goes, there is no reason to think that Jupiter will be more habitable then than it is now. The difficulty of the small supply of light and heat received from the Sun would apparently still remain, if indeed, the cooling of the Sun itself would not increase it. We do not know of any means by which our Sun could so increase its radiation as to supply to Jupiter from 25 to 30 times as much heat as it now receives, and this would be necessary to place it in the same favoured condition as the Earth. If so great a change were to take place in the Sun, life would be scorched out of existence on all planets nearer than Jupiter, and, similarly, if the solar emission were increased to meet the necessities of Uranus or Neptune, even Jupiter would fall a victim.

But we may consider it as a conceivable case that a planet of the exact dimensions of Jupiter may be revolving in an annual period of the same length as his, round some star that is capable of affording it adequate nourishment; and so with the three other giant planets. The actual Jupiter and Saturn of the solar system have, so far as we can tell, neither present nor future as habitable worlds, but we can consider what would be the case of imaginary bodies of similar dimensions in systems where the supply of heat would be sufficient. Or we can neglect the question of temperature altogether, as we did at first in the case of Mars.

All the four planets must shrink much in volume before their solidification will take place. Their average density at present but little exceeds that of water; indeed, Saturn is not so dense as water;

yet we must suppose that the same elements are in general common to the Earth and to them all. If we assume, then, that the four planets all cool to the point of solidification, their densities must be much increased, and their volumes correspondingly diminished. Since all four greatly exceed the Earth in mass, it is but natural to expect that, when they have assumed the terrestrial condition, they will be more closely compacted than the Earth, and their densities in consequence will be greater. It will, however, be simpler if we assume exactly the same density for them as for the Earth. Jupiter will then have shrunk to about one-fourth of its present volume, and the statistics for the four planets will run as in the following Table:

Statistics of the Four Outer Planets if with the Same Density as the Earth

Proportions of the Planets:—					
		Uranus	Neptune	Saturn	Jupiter
Diameter in miles		19300	20400	36000	54000
do	⊕ = 1	2·44	2·57	4·56	6·82
Surface,	⊕ = 1	6·0	6·6	20·8	46·6
Mass and Volume,	⊕ = 1	14·6	17·0	94·8	317·7
Gravity at surface,	⊕ = 1	2·44	2·57	4·56	6·82
Rate of Fall, Feet in the First Second		39·2	41·3	73·3	109·7
Atmosphere, assuming the total mass of the atmosphere to be proportional to the mass of the planet:—					
Pressure at the surface in lb. per square inch		88·2	97·0	305·8	685·0
Pressure at the surface in "atmospheres"		6·0	6·6	20·8	46·6
Level of half-pressure in miles		1·37	1·30	0·73	0·49
Boiling point of water at surface		127°C	129°C	148°C	164°C

Jupiter offers two peculiarities. In its shrunken condition, its diameter, instead of being eleven times that of the Earth, will be not quite seven, and the force of gravity at the surface will be greater than that of the Earth in the same proportion. A man who here weighs 150 lb. will there weigh over 1000 lb.; and the muscular effort of movement will be increased in the same ratio. The athlete who here can clear a height 5 ft. 8 in. will there, with like pains, surmount 10 inches; and other efforts will be in the same proportion. The atmosphere, supposing it to be in proportion to the mass of Jupiter, will exercise a pressure of 46½ "atmospheres," or more than 680 lb., to the square inch. Following on this enormous pressure at the surface would be the rapidity with which the atmosphere would thin out in the upward direction. The level of half-pressure would be attained by ascending less than half a mile in height; that is to say, there would be a difference of pressure of 340 lb. on the square inch from that experienced at the sea-level. We know from the fact that fishes live at enormous depths in the ocean, that living organisms can be constructed to endure great pressures, but they are not constructed to endure great alterations of pressure. The deep-sea fishes are as instantly killed by being brought up to the surface, as the surface fishes or the land animals would be if they were plunged into the depths. And it is clear that on Jupiter a low range of hills that on the Earth would be considered only an easy climb, would be an impassable barrier, not only from the immense exertion of mounting it, but chiefly from the unendurable change of pressure which the ascent would involve.

The sevenfold gravity of Jupiter, taken in connection with this enormous atmospheric pressure, would tend to make the meteorological disturbances of the planet violent far beyond anything of which the Earth can furnish an example. The atmosphere would possess a high viscosity, and differences in condition, pressure and saturation would tend to accumulate, until at length the balance would be restored with explosive suddenness and force. Here our most violent tornadoes may reach a speed of 100 miles an hour; on Jupiter, gales of five or six times that velocity would be common. We cannot conceive that living organisms would be able to grow, flourish and multiply where the conditions were so cataclysmic.

This difficulty must always exist where the planet is great in mass, and the force of gravity high at the surface. The case of Saturn is not so extreme as that of Jupiter, though it is probably sufficiently severe to exclude it from the ranks of worlds that could ever be dwelt in. The atmospheric pressure would be about 21 "atmospheres," or more than 300 lb. on the square inch. The level of half-pressure would be reached at about three-quarters of a mile, and the force of gravity be nearly 4½ times that of the Earth.

But the serious condition for Saturn would come from that feature which renders it by far the most attractive of all the planets seen in the telescope, the presence of the wonderful Ring system.

To us, viewing Saturn from afar, and from practically the same direction as the Sun, the Rings are seen lit up; but to a dweller on Saturn, the Rings during the day are between his world and the Sun, and hence turn their dark side toward him. More than that, the telescope shows us that the Rings cast a shadow on the planet; in other words, they eclipse part of it; and this shadow changes its position with the progress of the Saturnian year. Proctor computed that if the Rings were a hundred miles in thickness, the equator would suffer, in consequence, total eclipse for nearly ten days at each equinox, and partial eclipse for about forty days more. Moving away from the equator, each higher latitude would have a longer and longer period of eclipse in the winter half of its year; the higher the latitude, the later after the autumnal equinox the eclipse would begin, and the longer it would last, until about latitude 40° was reached. Here the eclipses would begin nearly three terrestrial years after the time of the autumnal equinox. At first the Sun would be eclipsed only in the morning and evening of each day, but the length of the daily eclipse would increase, until the Sun was hidden the whole day long. This period of total eclipse would last for about 6 years 8 months, terrestrial reckoning, or with the periods of partial eclipse, 8 years and nearly 10 months. Whatever the efficiency of the Sun that afforded light and heat to such a planet, it is clear that such eclipses must be fatal to life in two ways: light and heat would be cut off from wide regions of the planet for long periods of time, and terrible meteorological convulsions must follow in the train. Here on the Earth, though a total eclipse generally lasts only two or three minutes, the atmospheric disturbance is perceptible, and the fall of temperature very marked, and it does not require much reflection to see that the analogous disturbance in an atmosphere twenty times as dense must be terrific indeed during an eclipse that lasts not a few minutes only, but for more than six of our years.

The case of Uranus introduces us to another class of conditions fatal to habitability. The equator of Jupiter is inclined only 3° to the plane of its orbit; the difference in its seasons is, therefore, almost imperceptible; there is hardly any alteration in the incidence of the solar rays; it is, as if on the Earth, the height of the Sun at noon in mid-winter were what it actually is on the 14th of March, and its height at midsummer the same as we observe on March 28. The inclination of the equator of Saturn is considerably greater than that of Mars or the Earth, so that its seasons are more pronounced, but not to an extent that would introduce any radical difference. But for Uranus, the inclination of the equator to the plane of the orbit is 82°. If this were the case for the Earth, the noonday sun for London would be, at the spring equinox, 38½° high as at present, but its altitude day by day would increase with great rapidity, and before the end of April, the Sun at noon would be right in the zenith, and 13° above the horizon at midnight. At midsummer, indeed, it would be only 59° high at noonday, but it would be north of the zenith instead of south, and at technical midnight, it would still be 44° in altitude, thus moving round in a very small circle, only 15° in diameter. From about April 18 to August 25—that is to say, for 129 days—the Sun would never set, and unlike the summer day of our own polar regions now, wherein the Sun, though always present, is always low down in the sky, for much of that period it would pass the meridian quite close to the zenith.

As the year of Uranus is 84 times the length of our year, the London of Uranus would have to endure not far short of 30 years continuous scorching.

And the winter would be as long; the perpetual day of summer would be replaced by a night as enduring. More than 29 years of unbroken darkness, of unmitigated cold, cannot possibly ever consist with the conditions necessary for life upon a planet. Whatever the brightness of the imagined sun of Uranus, if for 29 years at a time that sun were below the horizon, the water on the planet must be congealed, and during the 29 years of unbroken day all the water would be as certainly evaporated.

Thus, though Uranus is not burdened by the enormous mass of Jupiter, nor overshadowed, like Saturn, by a system of rings, the extraordinary inclination of its axis introduces a condition which is as fatal to it, as a world to dwell in, as any of the disabilities of the other planets.

It is curious that these four outer planets, that resemble each other so strikingly in many of their conditions—in their vast size, high albedo, low density, and vaporous envelopes, that show, in their spectra, not merely the lines of reflected sunlight, but also special lines due to their own atmospheres (the chief of these being common to all the four planets)—should yet, in the inclination of their axes to the plane of their orbits, display every possible variety. The axis of Jupiter is almost normal to its orbit, that of Uranus lies almost in the plane of its orbit. The axes of Saturn and Neptune have a mean inclination, but it would appear that the rotation of Neptune is in the reverse direction to that of planets in general, so that the true inclination is usually taken as being the complement of the observed angle, as if the axis were turned right over. It is uncertain whether this would have any important effect upon the habitability of the planet, but it supplies the fourth possible case for the position of the axis.

CHAPTER XII

THE FINAL QUESTION

In passing in review the various members of the solar system, it has been seen that there are many conditions that have to be fulfilled before a planet can be regarded as the possible abode of life, because there are many conditions necessary in order that water may exist on its surface in the liquid state. The size and mass of the planet are restricted within quite narrow limits; and a world much larger or much smaller than our own is necessarily excluded. The supply of light and heat received from the Sun must not fall much below that received by the Earth, nor greatly exceed it; in other words, the distance of the planet from its Sun is somewhat precisely fixed, since the light and heat vary inversely not as the distance, but as its square. Of course, in different systems, with suns of different power, the most favourable distance will not be the same in each; but in any system there will be one most advantageous distance, and no great departure from it will be possible. This condition further implies that the planetary orbits must be nearly circular; pronounced eccentricity, such as the orbits of even our short-period comets display, would be fatal to the persistence of water in the liquid state, and hence to the continuance of life. A wide discordance between the planes of the planet's equator and of its orbit, by rendering the seasons extravagantly diverse, would act as prejudicially as an eccentric orbit, and a rotation period equal to that of revolution would mean that one hemisphere was eternally frozen while the other was exposed to perpetual heat.

It follows that in any given system there can be at most only one or two planets upon which life can find a home, and this only where the right conditions of size and mass, of rotation period, inclination of axis, and shape of orbit, all co-exist in a globe at the proper distance. But the type of system offered by our Sun and his planets is not the only one that exists. A very large proportion of stars are binaries—two suns revolve round their common centre of gravity. In many cases the two suns are separable in the telescope, and their relative movements can be measured; in other cases, termed "spectroscopic binaries," we only learn that a star which appears absolutely single has two components from the evidence of its spectrum; the spectroscope revealing two sets of lines that vibrate to and fro with respect to each other. Yet, again, a third class of double stars has made itself known in the "Algol variables." The optical double stars are cases where the two components are far distant from each other, and hence can be distinguished in our telescopes as separate points of light. The "spectroscopic binaries" are cases where the two components are too close to be separately perceived, but where the two are not greatly unequal in brightness, so that the spectrum of the one does not overpower that of the other. The "Algol variables" are cases where the two components are of very unequal brightness, and, being very close to each other, are so placed with respect to the Earth that the fainter partly eclipses the brighter in its revolution round it, and so causes a temporary diminution in its light at regular intervals. All these three classes of binary systems are now known to be very numerous. Prof. Campbell estimates that fully one star in six is a spectroscopic binary. But there must be many binary systems that do not reveal themselves—double stars where the companion is too faint or too close to be detected, Algol systems where the companion does not pass before its primary—and it seems almost certain that simple systems, like that of which our Sun is the unchallenged autocrat, must be comparatively rare.

But the problem of the movements of a planet attendant upon two or more suns is one of amazing complexity, and our greatest mathematicians have as yet only been able to deal with the approximate solution of a few very special cases. These are,however, sufficient to show that the orbit of a planet so placed would be most irregular; the variations in the supplies of light and heat received would be as great as even comets experience within the solar system, and, what would be more disastrous still, these variations would not be periodic but irregular. One year would be unlike that which preceded it, and would be followed by changed conditions in the next. Plants and animals would never have the chance of acclimatizing themselves to these ever-changing vicissitudes. The stability of condition essential for the maintenance of water in a liquid state would be wanting; and, in consequence, Life could neither come into existence, nor persist if it once appeared.

So far, therefore, our line of thought has led us to recognize that Life can exist in comparatively few of the innumerable stellar systems strewn through infinite space, and in any given system it can at best find only one or two homes. The conditions for a Life-bearing planet are thus both numerous and stringent—there is no elasticity about them. It is not sufficient that a planet might fulfil many or even most of these conditions; failure in one is failure altogether; "one black ball excludes;" the candidate who fails in a single subject is "ploughed" without mercy. And in most cases the failure is final; no opportunity is given to the candidate to "sit" again.

But Space is not the only horizon along which our thought must be directed; there is also the horizon of Time. Every world must have its Past and its Future, as well as its Present. For some worlds the conditions are so fixed that, like Jupiter and Saturn, they are not now worlds that can be dwelt in, they never were in that condition, and they never can be; their enormous mass forbids it. Mercury and the Moon at the other end of the planetary scale are also permanently disabled; their insignificant size excludes them. There was also a time when the Earth was not a world of habitation; it was "without form and void"; hot and vaporous, even as the four outer planets are now. Now it is inhabited, but there may come a time when this phase of its history has run its course, and either from a falling off in the tribute of light and heat rendered to it by the Sun, or from the gradual desiccation of the surface, or, perchance, from the slow loss of its atmosphere, it may approach the condition of Mars, and in its turn be no longer an abode of life. Many planets are essentially debarred from ever entering on the vital stage; but of those to which such a stage is possible, it can only form an incident in the entire duration of the orb. And if our Earth is any type or example of the vital stage in general, vast aeons must run their course from the first appearance of the humblest germs of life up to the bringing forth of Life in conscious Intelligence. One hundred million years are freely spoken of in this connection by those who study the crust of the Earth and those who are occupied with the relations of the varied forms of life. Man is the latest arrival on this planet, and however far back we try to push the time of his earliest appearance, it is beyond question that that time, relatively to the entire duration of the Earth since a solid crust began to form, is but as yesterday. If, from some other globe in the depths of space, this world of ours could have been watched during the long aeons that elapsed from its first separation from the solar nebula down to the time when it first possessed a surface of land and water, and from that time, again, throughout the hypothetical one hundred million years that preceded the advent of man, then, during all those aeons, those imagined observers would have had under their scrutiny a world as yet without inhabitant. The Earth now is in the inhabited condition, but science gives us no clue as to how long that condition will endure; rather such hints as are afforded us would seem to point to its lasting but for a brief season as compared with the indefinite duration which preceded it, and the indefinite duration which shall follow.

If this thought be sound, it places before us an entirely new and most serious consideration. The world predestined for habitation must not only have its size within certain narrow limits, its distance from its central sun in a certain narrow zone, its rotation period, the inclination of its axis, the eccentricity of its orbit, all suitable alike, but even if in these and in all other necessaries it is perfectly adapted for habitation, yet it will be only during a relatively small fraction of its entire duration that Intelligent Life, clothed in material form, will find a place upon it.

Let us sum shortly what we know and what we conclude. We know that this, our Earth, is a habitable globe, for we ourselves are living upon it. We know what constitutes the physical basis of our life, and under what conditions on this Earth it flourishes, and under what conditions it is destroyed. If we turn our eyes from this, our Earth, and look out upon the starry skies, we see the other planets of our system, and the suns which are the centres of other systems. From the consideration of the planets in our own system, we have seen how stringent and how many are the conditions imposed for Life to be possible. Round our Sun there is but a narrow zone in which a habitable world may circle; in this zone there is room for but few worlds, and we actually know of three alone, the Earth, the Moon, and Venus. We know that the Earth can be and is inhabited; that the Moon is not and cannot be inhabited; and that Venus, though of habitable size, may yet be subject to the fatal disqualification of always turning the same face to the Sun. Of other planetary systems than our own, we actually know of none, but we assume that there are such, and as numerous as there are suns in the starry depths. But of these planetary systems we can rule out, as containing no habitable member, all such as circle round double or multiple suns or, indeed, round any single star that, from whatever cause, is largely variable and, therefore, much less stable than our own. Mira Ceti, which in 5 months increases its brightness 1000 times, may stand as an example. Probably these disqualifications rule out of court the great proportion of the stellar systems. Of the few, comparatively speaking, single and stable suns that remain in the heavenly abyss, we must conclude, from what we know of our solar system, that they, too, have but a narrow zone, outside of which no world would be fit to dwell in; whilst in the zone the few worlds which might exist must violate no one of many strict conditions. If we assume that there are a hundred million stars within the ken of our telescopes, we may well believe that not more than one in a hundred of these would fulfil the condition of being a single and stable sun, such as ours. Of the planets revolving round these million suns—stable and efficient suns—can we expect that in more cases than one in a hundred there will be a planet in the habitable zone fulfilling all the other conditions of habitability, of size, mass, inclination of axis, circular orbit, and rotation? Of these ten thousand earths which may be made fit for the habitation of Man, can we assume that even one in a hundred is now at that epoch in its history when it is no longer "without form and void," when a division has been made between the waters under the firmament and those that are above the firmament; when the waters under the heaven have been gathered into one place, and the dry land has appeared, and when the earth and the waters have brought forth life abundantly? Out of a hundred million of planetary systems throughout the depths of space, can we suppose that there are even one hundred worlds that are actually inhabited at the present moment? These numbers and proportions certainly are not, and cannot be, based on knowledge; they are given as illustrations only; but, vague as they are, they suggest that our Earth may be neither one of many inhabited earths, nor yet unique, but one of a few—indeed of a very few.

And then the objection is raised: "If our own Earth is but one of, perhaps, two inhabited worlds in the solar system; and of perhaps one or two hundred inhabited worlds throughout the furthest space that we can scan; why is all this waste?" Of all the countless millions of stellar systems without living organisms as inhabitants, we cannot tell the purpose for the simple reason that we do not know it; but of "waste" in the solar system, there is no question. Relatively speaking, this is quite insignificant, for we cannot consider that as "waste material" which is useful and, indeed, essential to existence. For, consider first the material in the Earth itself. Its total volume is 260,613,000,000 cubic miles, but man only lives *upon* its surface of less than 200 million square miles in extent, and he can not probe down as far as ten miles below it, through the depths of ocean or by his deepest mine. Thus we are left with over 258 thousand million of cubic miles that man, or plant, or beast can never make direct use of. But without this 258 thousand million cubic miles that he can never sow nor reap, the overlying platform on which he dwells would be useless for retaining the air or the water by which he lives. No less essential is the Sun; its vast bulk of

2,000,000,000,000,000,000,000,000,000 tons

can, in no single unit, be counted "waste," for it is from this that the heat and light necessary for life on the Earth is derived. But the tonnage of all the planets combined is but 0·13 per cent of the Sun alone; and a wastage, if such it is, like this is insignificant from a material point of view.

There is a type of politician at the present day who is convinced that the highest purpose to which land can be put is to build upon it; that being, in general, the use giving the highest money return per square foot, though the return does not always fall to the builder. It has taken not a little agitation and popular pressure to enforce the truth that cultivated land is also of use. But there are few who realize that land that is neither built upon nor cultivated is also essential. Our barren moors and bleak hillsides, "wastelands" as we call them, are absolutely necessary as collectors of the water by which we live. From them our springs take their source; and they supply our cities with the first necessity of life.

We find, then, in this universe so far as we can know it, that Space is lavishly provided, Matter is lavishly scattered, Time is unsparingly drawn upon, but Life in any form, and especially in its highest form, is, relatively speaking, very sparsely given. That very circumstance surely points to the overwhelming importance of conscious, intelligent Life, and the insignificance of lifeless matter in comparison with it. We have to exhaust arithmetic in computing the size, the mass, the output of heat and light of our Sun, yet it is but the hearth-fire and lamp of terrestrial life; and its amazing agglomeration of matter and energy is ungrudgingly devoted to this humble purpose. Whatever view we hold as to the scheme of the universe; whether with the unthinking we fail to recognize Thought and Purpose behind its marvellous manifestations, or, with the thoughtful, realize that only Infinite Thought could provide so wonderfully for the bringing forth of thought in living material organisms, the conclusion still remains: living intelligences are, by the direct testimony of the universe itself, its noblest and most precious product.

The plea is often made that as we find life adapting itself to a great variety of conditions on this Earth, we must not set limits to its power of adaption to the conditions of other worlds. But this plea is an unthinking one. The range of conditions through which we find life on this Earth is as nothing to the range given by the varied sizes and positions of the different planets; and even on our Earth, life in the unfavoured regions—the tops of mountains, the polar snows, the waterless deserts, the ocean depths—is only possible because there are more favoured regions close at hand, and there are, as it were, "crumbs that fall from the rich man's table." A well-known littérateur in setting forth "a hundred ways of making money" gave great prominence to the method of living as caretaker in an empty house. But residing in an empty house does not, in itself, supply the means of sustenance; these have to be furnished by the wealthier man who employs the caretaker.

Another plea for vague sentiment in this matter is that we cannot expect that intelligent beings on other worlds would have the same form as man, and if not the same form, then, that the same conditions of existence would not hold good for them as for us. Both contentions are unsound. Protoplasm is the physical basis of all the life that we know, whatever its form; though these forms are to be counted by the million, and are as diverse as they are numerous. And everywhere and always, water is found essential to protoplasmic life. Of life of any other kind we do not know any examples; we have no instance; if such exist, then they are beyond our ken.

And neither anthropologist nor biologist would admit that the form of intelligent life was an unrelated accident. Whether the form brought the intelligence, or the intelligence the form, or both were evolved together, the one reacting on the other, the human form and the human intelligence are associated, and we feel this to be so of necessity. In 1891, Dr. Eugene Dubois found in Java a molar tooth and a portion of a skull, and later the thigh bone of the left leg, and two more teeth. Such as they were, these relics appeared nearer in form to the corresponding fragments of an average Australian than to those of an ape, and on this ground intelligence was claimed for the creature of which they were the remains, and it was given the name of Pithecanthropus, or Ape-Man. The discovery aroused much discussion, but on all sides it was unhesitatingly assumed that the difference between the form of Pithecanthropus and that of the most similar ape was an index of its superior intelligence over the ape, just in so far as that difference was in the direction of the modern human form. The same remark applies to the recent discovery of very ancient human remains in Sussex. Never at any time has it been supposed that the physical frame has followed any other path in the evolution of intelligence than that which brought forth man. The flesh-eating animals have attained efficiency in hunting and warfare by variation along many types of form; the herbivora have been not less varied in the forms by which as races they secured themselves from destruction; but Thought has been associated with the development of one type or form only, and the entire future of Thought on this planet rested neither with mammoth nor cave-bear, but with the possessor of the erect stature, the upward look, the differentiation of hand and foot, even in their crudest and earliest stages.

Swift, in *Gulliver's Travels*, conceived of a land where the intelligence and conscience of Man dwelt in the form of the horse, and the human form tabernacled the instincts of the beast. H. G. Wells, in his *War of the Worlds*, attributed intelligence to monsters—half-cuttlefish and half-anemone,—and the human form to their helpless, unresisting prey. Both conceptions are as scientifically absurd as they are gross and revolting; and if it were possible for the skeleton of creatures from other worlds to be brought to us here, then biologists would as confidently pronounce on their intelligence as they do on the extinct forms of bygone ages—the nearer to the human form, the nearer to the human mind. We have found the figures of reindeer, horse, and mammoth scratched in outline on a mammoth tusk; but though the artist has left no other trace, we need no further evidence of his bodily form. Neither horse, nor reindeer, nor mammoth made those rough outlines; they were drawn by a man. More striking still, France yields us chipped flints by the million, flints so slightly shaped that it is in dispute whether they may not have been so broken by the action of torrents. But there are only two theories about them; either they were so chipped by natural action, or they were designedly so chipped by creatures resembling ourselves in head and hand.

The question that has been dealt with in this volume is a scientific one, and the attempt has been made to treat it as such, and to argue from known physical facts as to the conditions of worlds which we cannot visit. But by many the question is generally discussed wholly apart from physical facts at all, and it becomes one of sentiment and of religious sympathy. Yet, curiously enough, the division between those who think that all worlds must be inhabited and those who think that our own world stands alone is not coincident with any line of theological divisions, but rather cuts across all such. Some believers in Christianity argue that since God has filled this world with Life, Life has been His purpose in the world, and must therefore have been His purpose in all other worlds—they too must be filled with Life in like manner. Other believers argue that this world was the scene of the Incarnation of Our Lord, and is therefore unique in that respect; and that this uniqueness sets its stamp upon this world in all respects. Opponents to Christianity are divided into the same two classes, the one arguing that wherever there is matter the inevitable course of evolution will produce life, and eventually intelligent life. The other class are equally clear that all forms of life are special, the result of the particular environment, and that it is unreasonable to expect that any other world has had the same history as our own, or that the same special conditions have prevailed elsewhere. In other words the belief that there are other inhabited worlds has depended chiefly neither on science nor on religious belief, but upon sentiment. There are some who like to think themselves, and the race to which they belong, altogether exceptional; others delight in finding themselves reflected wherever they look. So far as Science has progressed and can return an answer to an enquiry that exceeds so far the bounds of our direct observation, it dissents from both orders of thought. The conditions of life are indeed narrow, special, restricted; intelligent, organic life must, relatively speaking, be a rarity in the universe, but we lack the information that would enable us to affirm with any confidence that such life is only to be found upon this world of ours. Heavy as the odds are against any particular world being an inhabited one, yet when the limitless extent of space is considered, and the innumerable numbers of stars and systems of stars, it seems but reasonable to conclude that though inhabited worlds are relatively rare, the absolute number of them may be considerable; considerable, if not at one particular moment of time, yet when the whole duration of the universe is admitted.

But there is a religious question connected with this enquiry; one that goes down to the very roots of man's deepest thoughts and aspirations. As individuals our days on the Earth are as a shadow, and there is none abiding; as individuals we pass and disappear; and though the race remains, yet as far as science can guide us and enable us to penetrate the future, the same lot awaits the race as well. Slowly but surely the water of a planet will combine with its substance or disappear into its crust. The cooling of the Sun, though it may be long delayed, would seem to be inevitable in the sequel.

"Oh, life as futile then as frail.
....
What hope of answer or redress?
Behind the veil, behind the veil."

It is to this veil that we are now brought. It seems impossible to believe that Life, so rare a fruit of the universe, intelligent Life, conscious Life, to which the long course of evolution has been so manifestly leading up all through the long ages, should have[Pg 160] no better destiny than a final and hopeless extinction; that this Earth and all the efforts and aspirations of the long generations of men should have no worthier end than to swing, throughout the eternal ages, an empty, frozen heap of dust, circling round the extinct cinder that was once its Sun. If we look backward, we seem to discern clear signs of progress; if we look forward, we discern nothing but the veil. Science is but organized experience, and experience of the future we have none.

There was a time when on this world there was no life; a time when life began. How did it begin? Under what conditions?

Of that great change—when non-living matter first became endowed with life, became so endowed not by the action and intervention of other living matter, but without it—we have no knowledge, no experience. And so long as this continues to be the case, that change, the greatest physical change that has yet taken place in the history of the universe, the first change of the non-living into the living, is outside the reach of science; it lies beyond its border. We may guess and speculate about it, but speculation is not science; we may spin words about it with the utmost skill of the dialectician, but metaphysics is not science; it can never come within the scope of science until it has first come within the scope of experience.

There is, therefore, a veil behind us as well as the one that encloses us in front; and as hitherto Science has failed to pierce the veil of the past, it is even less able to pierce the veil of the future; for of the future we have no experience.

Here, then, our enquiry must end, for it is an enquiry of physical science; the search for living material organisms endowed with intelligence. How life first came upon this Earth, or when, or where, is beyond the power of science to determine. Yet it did come. There was a time when there was no life here; none, not even the humblest form of it; nor was there any hint or foreshadowing of it, still less of all its infinities of form, and possibilities of development.

Once Life was not, yet Life came, and now, life is abundant, but abundant only in worlds quite exceptional in their conditions, and therefore few in number; it is even conceivable that this Earth of ours may be unique. But life as we know it, protoplasmic life, life dependent upon water, the life of intelligence united to the material organism, is under sentence of death. Has it any future beyond that veil? Is there any kind of life not subject to these narrow limitations; not under the inexorable decree?

To questions such as these Science has no reply to give; it is even more helpless to answer them than to determine how life first came; its experience does not reach so far. Science can examine the present conditions of physical life, but whether or no that life can undergo a change greater than that which passed upon the old inorganic world, it cannot determine. It has no experience.

But if Science is dumb, if the utmost exertion of human energy and power of research can throw no light on a future of which we have no experience, we are not left without an answer. A voice has been heard, the voice of the Son of God Himself:

"I am the Resurrection and the Life. He that believeth on Me, though he were dead, yet shall he live."

And accepting His word, the Church in all ages, and among all nations, peoples, and tongues, has made reply:

"I look for the resurrection of the dead and the life of the world to come."